In the Shadow of the Moon

IN THE SHADOW OF
THE MOON

THE SCIENCE, MAGIC,
AND MYSTERY OF
SOLAR ECLIPSES

● ● ●

ANTHONY AVENI

Yale
UNIVERSITY PRESS

New Haven and London

Published with assistance from the foundation established in memory of
Philip Hamilton McMillan of the Class of 1894, Yale College.

Yale University Press books may be purchased in quantity for educational,
business, or promotional use. For information, please e-mail sales.press@
yale.edu (U.S. office) or sales@yaleup.co.uk (U.K. office).

Set in Gotham and Adobe Garamond type by IDS Infotech, Ltd.
Printed in the United States of America.

Library of Congress Control Number: 2016953812
ISBN 978-0-300-22319-4 (hardcover : alk. paper)

A catalogue record for this book is available from the British Library.

This paper meets the requirements of ANSI/NISO Z39.48–1992
(Permanence of Paper).

10 9 8 7 6 5 4 3 2 1

To Ed Maxwell—wise thinker, new friend

Contents

CONTENTS

TOTALITY

ECLIPSES IN THE MODERN AGE

CONTACT THREE

LESSONS FROM ECLIPSES

CONTACT FOUR

PERSONAL ECLIPSES

Preface

Trained in the sciences, I was once too engrossed in my specialty, the astrophysics of star formation, to think much about how different cultures might interpret the cosmos. Then I took a group of students to Mexico to measure the celestial orientation of pyramids at Teotihuacan and Chichén Itzá.

By chance, I read reports in the literature about suspected astronomical alignments in sacred architecture, especially among cultures like the ancient Maya of Yucatán, and I had developed a method of precise surveying using celestial referents that caught the eyes of archaeologists. They invited me to try my techniques, and in return they taught me about the ancient peoples who watched the sky. We created a new interdisciplinary field variously called *astroarchaeology*, *archaeoastronomy*, and later (and less intimidatingly) *cultural astronomy*.

This fieldwork at the ancient Maya ruins in the rain forests of Central America opened my eyes to a sophisticated culture skilled in mathematics, writing, and the precise observation of nature. Our most recent work on the decipherment of a text written on a wall at the recently excavated city of Xultun in the Guatemalan rainforest proves that Maya astronomers, motivated by interests quite different from what *we* call science, were capable of predicting eclipses hundreds of years in advance, with minimal technology.

Modern astronomers' questions about their ancient predecessors focus largely on aspects of the *things* in the sky: Did they mark sunrises and sunsets? What did they observe about the moon besides its phases? Did they predict eclipses? Did they know the world was round, that the sun is the center of the solar system, that we're part of the Milky Way Galaxy in an expanding universe filled with galaxies like it? In other words, how like *us* were they? You could pose the same question about extraterrestrials.

Once I began working with archaeologists and anthropologists, I got exposed to questions of a different sort: What does another culture's astronomy tell us about its people's religion, their beliefs in an afterlife, the way they treat their dead? Did they practice astrology? To what end? Did the ability to predict what was about to happen in the sky over their city affect their politics? What was their concept of history? Did they believe in their cosmic myths? Such questions center around what the sky meant to *them.*

Cultural astronomers are more interested in what people believe about celestial happenings than the happenings themselves. Whether modern Western civilization might label these beliefs religious, astrological, superstitious, or nonscientific, understanding *their* perspective allowed me to hold up a mirror to my own acquired scientific view of nature as a world where things happen without regard to human affairs. I see my face in that mirror, set against a background of other faces. The minds behind them harbor alternative ways of finding meaning that make sense to them. Drawing constellations or star patterns, following the movement of planets across a zodiac, or being dazzled by a solar eclipse—experiencing the sky is not the same for everyone. As an astronomer and an anthropologist, I find most explanations about how people around the world understand natural phenomena a bit short-sighted. This book represents my attempt to broaden our collective vision. There is much to be learned, shared, and felt about the human condition from a deeper exploration of the natural world seen through diverse human eyes.

In the Shadow of the Moon is a personal narrative that tells as much about people who have watched eclipses as the eclipses themselves. I was motivated to craft this book for publication in 2017 because, after experiencing a ninety-nine-year eclipse drought, mainland North America is on the verge of witnessing two transcontinental total eclipses of the sun. The twin eclipses positioned just over the American horizon offer an ideal prompt. If two minutes of dramatic daytime darkness generated more than one hundred articles in the *New York Times* in January 1925, imagine the attention that will attend seven minutes of totality spread seven years apart in large swaths centered on the middle of North America between 2017 and 2024.

I've been the fortunate witness of eight total eclipses of the sun, bringing me to diverse places ranging from Canada to the South Pacific, across Atlantic waters off the coast of Argentina, to the Egypt-Libya border and the middle of the Black Sea. Despite all this travel, I've logged less than half an hour in the complete darkness of the shadow of the moon. This is far from record breaking, and any amount of time is never enough. If you've already witnessed a total solar eclipse, you know what I mean.

CONTACT ONE
An Introduction to Solar Eclipses

• • •

(*Overleaf*): Contact I / Begin Partial (© 2008 Fred Espenak, MrEclipse.com)

1

Colossal Celestial Spectacles

These late eclipses in the sun and moon portend no good to us:
though the wisdom of nature can reason it thus and thus, yet nature
finds itself scourged by the sequent effects. Love cools, friendship
falls off, brothers divide: in cities, mutinies; in countries, discord; in
palaces, treason; and the bond cracked between son and father. This
villain of mine comes under the prediction; there's son against
father: the king falls from bias of nature . . . 'Tis strange!

—*William Shakespeare*, King Lear, *1606*

Studying the sky has made me realize what sheltered lives we lead.
Efficient central heating and air conditioning signal our preference
for being indoors. Most of us labor indoors, shop and dine indoors,
work out and play tennis indoors. Our eyes spend most of their
waking hours scanning computer and phone screens. We tan without
exposing ourselves to the sun and, mindless of its daily course, we
look instead at our watches to tell time and set appointments elec-
tronically. Think of the inconvenience the last time you experienced a
power outage.

Living in such a world, we tend to dwell on the gains technology
has afforded us: instant news, fresh food, the ability to move rapidly
from place to place. But technology isn't the only reason we severed
the links between daily life and the physical world around us. A

change in the basic way we *think* about nature contributes to our detachment from it. The divorce from the world around us began five hundred years ago during the European Renaissance and continued through the late eighteenth-century Enlightenment, when nature was reintroduced to culture as a world apart from human affairs. Like the earth, the sky, once alive to our ancient ancestors, came to be regarded as inanimate matter, possessing neither soul nor meaning. Western society attempted to objectify the universe, making it an entity to be described and understood *as it is* or *for its own sake*. Consequently, whatever dialogue we once had with nature was silenced.

When we do encounter the world outside, we tend to cast our eyes downward, ignoring half our visual field, which lies above eye level, bounded by the horizon that surrounds us. But who *needs* the sky, anyway? What incentive is there to know the constellations, the phase of the moon, or the point where the sun popped up over the horizon this morning? Today the look of the sky is little more than an imminent weather indicator. Shall I take an umbrella? And when any of us contemplate a recreational peek at the heavens, we dig instead into our techno-devices, like the apps that allow us to track the stars from the comfort of a basement apartment.

But for most of human history, the sky *was* relevant. People paid attention to the rising and setting sun, the phases of the moon, the coming and going of each of the planets. The relative perfection of the firmament beckoned for human connection. The first crocus might arrive a bit late, the last snowfall a little early, but I know that when Arcturus, the brightest star in the northern hemisphere, makes its first annual appearance in the east after sunset, it's my birthday. The sky became the logical medium to mirror the ordered lives our species strove to lead. For ages it would serve as the storyboard for morally based tales of heroism and adventure. From season to season, people found meaning in the dance of the cosmic denizens who resided in the world above.

Every once in a while, something extraordinary happens up there, grabs our attention, and reconnects us. Comets belong to my personal colossal celestial spectacles, like the comets Ikeya-Seki in 1965, West in 1976, Hyakutake in 1996, and Hale-Bopp in 1997. (Halley's Comet was a bust in 1986, but on its previous close passages in 1835 and 1910, it took over the heavens and the headlines for weeks.)[1]

Most comets are unpredictable. Formed in the far reaches of the solar system, they appear suddenly seemingly out of nowhere, streaking slowly from night to night across the starry background, most of them never to return. That's probably why throughout history, comet appearances were thought to portend ill fortune, unscheduled events that disturbed the otherwise pristine, predictable happenings in the quintessence. Julius Caesar's assassination coincided with the appearance of a comet. The comet of 1316, visible across northern Europe, was said to have been a sign of the Great Famine of 1315–17, which killed 10 percent of the population. (The extreme weather known as the Medieval Warm Period was more likely responsible for the event.) The great comet of 1843, visible as a luminous streak across the western sky for several weeks following sunset, was heralded by a religious sect that would later grow into the Seventh-day Adventist Church as a "sign in heaven" that the Apocalypse described in the Book of Revelation was surely at hand. And when Halley's Comet appeared in 1301 (it wasn't yet known to be recurrent), Giotto imaginatively represented the Star of Bethlehem as a comet in a nativity painting on the ceiling of the Scrovegni Chapel in Padua, Italy.

How bright a comet will become (and when) is difficult to predict. It depends on its distance from the earth and how close it gets to the sun. Comets shed their gaseous shroud in the form of a tail that grows to millions of miles in length. Not all comets possess the same blend of dust and gas: lots of dust and you get a faint comet, less and you get a much brighter one due to the sun energizing the molecules that light up its gaseous atmosphere.

In 1965, as a young astronomy teacher in Hamilton, New York, I remember driving to the Colgate University Observatory for the anticipated pre-dawn view of Comet Ikeya-Seki. Turning onto the main campus road and veering toward the observatory, I noticed several pairs of headlights swarming over the dew-covered grass in the darkness. Other vehicles were parked along the crowded roadside all the way down the hill. A few dozen people milled around the tiny oracle housing our telescope, eagerly awaiting instructions about which way to direct their gaze. Some of my academic colleagues made skeptical inquiries, an attitude well cultivated in their trade. There were a couple of excited kids who aspired to be astronauts, growing up in the mid-1960s when America's space program was just beginning. Would the comet inspire them?

Though we persevered morning after morning for several days, we saw very little except for a faint fuzzy patch at the predicted position. Then, thinking the comet a dud, everyone gave up. About three weeks after predicted maximum, I awoke just before dawn with a start. Our four-year-old had toppled out of bed. Once I got up and repositioned her, I thought: why not take a look outside? I stepped out into the chilly autumn air, and there it was, an extraordinarily bright fuzzy object just above the faintly twilit eastern horizon. Ikeya sported a long, curved luminous tail that arced almost all the way up to the overhead point. Never had I seen nature conjure up a sight like that; it just didn't seem to belong in a starry sky. Ikeya was by any account a colossal celestial spectacle! To this day it remains the *second* best wonder I've ever witnessed in the sky. Unpredictably, the comet turned out to be more of a sun grazer than orbital pundits had anticipated. In fact, it passed so close to the sun that it was captured and locked into an elliptical orbit. It's due back in 2064.

In our scientific age, we've been taught rational explanations for rare cosmic events, yet we are still provoked by "nature's bias," as Shakespeare wrote. We are still surprised, awed, and maybe even a bit unsettled, as was I by Ikeya years ago. People still wonder: can there be a message that accompanies this fractious cosmic behavior?

Securing omens from heaven lives on in popular segments of American culture. I vividly recall Comet Hale-Bopp playing the role of extraterrestrial messenger three decades later on the eve of the third millennium. Leaders of the Heaven's Gate cult sought signs from above that the destruction of the world was imminent, at a time when thoughts about the end of the world in general were on the increase.[2] They believed Hale-Bopp was being trailed by a space vehicle inhabited by descendants of ancient astronauts who had originally placed our human ancestors on earth. Now they were returning to reap the harvest, those who had sacrificed themselves to get to the next step in the universal course of galactic civilization. Between March 24 and 26, 1997, thirty-nine members of the group donned identical shirts, sweatpants, and shoes. They drank a phenobarbitol-laced concoction to prepare for the anticipated cosmic boarding. Days later, police found their dead bodies. Comets can have a profound effect on people.

Meteor showers are high on my list of colossal cosmic events. The sun eventually destroys comets with closed orbits, particularly short-period ones. The disintegrated debris of these worn-out comets becomes seasonal storms of "shooting stars" when the earth crosses their orbits, the Quadrantids in early January, the Leonids in mid-November, the Geminids in mid-December. Most famous among the showers are the summertime Perseids, which dart in all directions out of a radiant point in the constellation of Perseus during mid-August. On a moonless night, especially after midnight, when our part of the world meets the swarm head on, you can count up to a hundred of these luminescent yellow-white streaks in just an hour, some up to dozens of moon-widths long, each lasting for a few seconds. I remember instinctively anticipating the "pop" of a Roman candle sound as each meteor trail faded.

The Leonids become especially active at thirty-three-year intervals, when the earth crosses the portion of the orbit where the defunct comet's remnants are most heavily concentrated. The 1833 storm rained down close to 30,000 meteors an hour. "The meteors fell from the elements on the 12th of November 1833 on Thursday in Washington. It frightened the people half to death," wrote one diarist.[3] Astronomy historian Agnes Clerke compared their intensity with what you might see during a mild snowstorm: "The sky scored in every direction with shining tracks and illuminated with majestic fireballs."[4] Frederick Douglass and Harriet Tubman witnessed and wrote about the great Leonid shower of 1833, as did the founder of the Mormons, Joseph Smith, who interpreted the storm as a sure sign in the sky that Jesus was about to descend from heaven.

Occasionally a super meteor, or bolide, will flash unpredictably out of nowhere. In Chelyabinsk, Russia, on February 15, 2013, a streaking flash lit up the morning sky brighter than the sun; it exploded near the horizon after a few seconds. The shockwave knocked out windows in several houses and buildings and left a sulfurous smell. The twenty-yard-diameter asteroid fragment responsible for the blast was estimated to weigh upward of 10,000 tons. You need to be in the right place and facing the right direction to witness a bolide.

Recently, while talking to one of my students on the campus quad following our early afternoon class, I saw his face light up. His jaw dropped as he pointed over my right shoulder. By the time I turned around, all I could see was a remnant smoke trail running from the top of the sky down toward the horizon. As I said, you need to be lucky.

The northern lights deserve a high place on anyone's list of nature's celestial extravaganzas. The dazzling shows they put on are semi-predictable, occurring most often a day or two after hot flares erupt on the sun's surface. Incoming solar electrons and protons collide with molecules and atoms at the top of our atmosphere, causing them to glow. Drawn along the earth's magnetic field lines toward the poles, the particles create a display of shifting, pulsating curtains of colored light: green and blue from activated nitrogen molecules, red from oxygen atoms (you can find a host of displays on YouTube).

Because I happen to live in a favored location relatively close to one of the magnetic poles, I've had several opportunities to witness the aurora borealis, the "northern dawn." I know I'm in for a show once I spot that whitish-green glow near the north horizon. Greenish bands develop into curtains that start to ripple as if blown by the breeze; then they break down into arcs, only to re-form. In one unusually strong display, I saw bright crimson rays emerge from a focus near the overhead position. They dissolved into pulsating arcs that seemed to shower down on me, the funneling of charged particles along the magnetic field lines. After half an hour the colors faded, but they became reactivated an hour or two later.

Contemporary Inuit people of northeast Canada call the northern lights *aqsarniit*, after the celestial football players who, they say, make it happen. An old Inuit man described the impression he had of the aurora in his youth in animate terms. He believed his actions could influence what happens in the heavens: "I have heard that they used a walrus head for a football. . . . I have not heard what the *aqsarniit* are made of but you can hear the swishing sound they make. If they get too close you can chase them away by twisting your tongue to the upright position. We used to do this when the *aqsarniit* got too close and it always had an effect. When one was watching them you could see them getting closer and closer until one could hear the swishing sound. At the same time they would be moving sideways also making this sound."[5] Scientific narratives of dazzling celestial phenomena like these usually include an abundance of detailed explanations of the physical causes behind the visual effects. Little attention gets paid to exactly what is *felt* by those who thrill to them.

Rainbows, halos around the sun and moon, lightning strikes in the distance that illuminate the night sky, comets, meteors, the aurora—all are high up on my personal list of colossal celestial spectacles. But of all the wonders of the sky I have ever marveled at, none came close to surpassing the transient, exquisite beauty of a total eclipse of the sun.

2

Watching People Watching Eclipses

Nothing can be surprising any more or impossible or miraculous, now that Zeus, father of the Olympians has made night out of noonday, hiding the bright sunlight, and . . . fear has come upon mankind. After this, men can believe anything, expect anything. Don't any of you be surprised in [the] future if land beasts change places with dolphins and go to live in their salty pastures, and get to like the sounding wave of the sea more than the land, while the dolphins prefer the mountains.

—*Greek poet Archilochus, seventh century BCE*

"Storm chasers" deliberately pursue severe weather conditions, especially tornadoes. The term dates back to the 1950s, though it didn't enter American pop culture until television programs focused on the scientific value of collecting data in the eye of a storm in the late 1970s and early 1980s. Later, its potential for extreme recreational activity surged. The 1995 box office hit *Twister,* which followed storm chasers-cum-meteorologists testing their competing devices during a rash of storms in Oklahoma, further romanticized the dangerous activity.

The desire to live on the edge, to get as close to danger as you possibly can simply for the thrill of it, compels adventure-seeking extroverts to engage in extreme sports, such as rock climbing, BASE jumping, and reaching the peak of Mount Everest. These same tendencies can be seen in storm chasers, motivated by the pull of nature—as one journalist put it, "to see something magnificent that you can't possibly control."[1] Recreational storm trackers love photo-documenting the biggest, the strongest, often the most damage-inflicting weather events. Risk-oriented travel companies offer "chase tour" services; not surprisingly, they attract people who live mostly outside the Tornado Alley states, including people from other countries. A typical tour will pick you up at the airport (in season, of course), supply you with ground transportation, lodging, food, snacks, video highlights, a T-shirt, and souvenirs. Participants tell stories of the tension they experience waiting out incipient danger and the devastating letdown that follows when they fail to sight a storm or, worse, chase one only to miss getting close enough to document it.

But storms aren't the only natural phenomena adventurers lust after. Eclipse enthusiasts abound. A popular eclipse website predicts: "People from all over the world [will] begin to converge on the United States [in August 2017]. Except for people returning home, visiting family, or conducting business at what happens to be just exactly the right time in history, these will be people who make it a point to travel to wherever the moon's shadow is going to touch the earth, and position themselves in a spot carefully chosen—sometimes years in advance—to ensure they see the sight."[2]

The people referred to are the eclipse chasers. They are after a different kind of thrill, and they have been around far longer than seekers of the storm. Eclipse chasers acquired their contemporary moniker from a 1914 *New York Times* article describing an early attempt by Amherst College astronomy professor David Todd to view totality from "an aeroplane," saying that he was "chasing the sun."[3]

Today, a host of websites unite the questers of the lunar shade; for example, the website Eclipse-Chasers.com calculates eclipse events

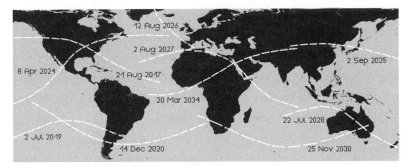

Projected paths showing locations to view total solar eclipses including the twenty-first-century great American eclipses. (Permission to reproduce from Exploratorium, www.exploratorium.edu. Basemap data © OpenStreetMap contributors [CC BY-SA].)

and logs observations from contributors. Bill Kramer, who created the site, is an engineer, a computer applications developer, and an avid amateur astronomer. (He is also a lecturer on astronomy-related theme cruises.) The site's "Eclipse Chaser Log" lists the number of eclipses experienced, totality time, partial eclipse time, and the chase success percentage of 220 entrants. At this writing, the record holder in this competitive cosmic sport is Williams College astronomer Jay Pasachoff, with sixty-four eclipses, thirty-two of which were total (tied with another entrant). Pasachoff has experienced $1^h19^m31^s$ of totality and a total of $143^h14^m28^s$ under the influence of the shadow, counting partial eclipses.[4] Extreme eclipse competitors try to prolong their totality time by literally chasing the eclipse, flying in the same direction as the shadow, which travels across the surface of the earth at nearly two thousand miles per hour. In 2010, along with twenty-three others, commodities trader Rick Brown chartered a jet that traveled five hundred miles per hour, allowing him to extend his time in the shade by four minutes.[5]

"I am an eclipse chaser and I travel all over the world. You may think I'm crazy. I prefer interesting," reads one T-shirt available on Eclipse-Chasers.com. The serious umbraphile thinks nothing of shelling out vast sums of money to bask for only a few seconds in totality. He usually gets hooked when a friend shows him pictures of her own experiences under the shadow and persuades him to join in her next venture. It also helps if you're the gregarious type with an appetite for exotic food.

"We always manage to talk ourselves into going no matter how expensive it is," said a couple who had just shelled out $8,000 apiece, plus airfare, for an eight-day trip to Easter Island in 2010 that would give them four and a half minutes in the shade.[6] Luxury tours for that event included a six-night cruise around the islands of Tahiti, with private lectures from eclipse experts, at a cost of up to $30,000. Some tour operators booked entire island facilities along the eclipse path. Clearly, eclipse chasing is accessible mostly to the wealthy.

Thrilling to the power of nature to overwhelm the senses is one motive. "That moment alone you feel like, 'yes, we did it.' All the natural power comes together when you see it and even though it's short, I am never disappointed," remarked a chaser, one among eight thousand who traveled to the Faroe Islands in 2015 for two minutes of daytime darkness. He added, "My family and some of my friends think I'm a bit crazy but every person goes crazy in their own way."[7] "It makes some people think about religion, or how small we are and have little control of what is going on around us," said another.[8] An eclipse chaser has made a life choice to reexperience the thrill and excitement of totality, wrote Kate Russo, an eight-eclipse veteran and author of *Total Addiction: The Life of an Eclipse Chaser.* Her motive, she tells readers, is part of an innate response to the fire ignited by the eclipse experience. "It's not just a hobby—eclipse chasing is a way of life."[9]

On the other hand, one person's thrill of a lifetime can be a big yawn to another. William Phelps, a Yale English professor and early twentieth-century eclipse chaser, tells the story of a friend's relative, a Harvard grad, who happened to be in a position to view the August 1932 eclipse. Allegedly, just before totality began, he remarked that this was his regular time to go to the bathroom—and disappeared.[10]

Ecstasy is a common feeling described by eclipse chasers, especially under the influence of more than the moon's shadow. Inspired by Buddy Holly's hit tune "Rave On," by the 1980s and 1990s a kind of psychedelic electronic dance experience known as "raves" attracted thousands (particularly young Europeans) to underground concerts in outdoor countryside locations. Ravers made use of pagan symbolism to create a back-to-nature fantasy world, and chasing eclipses became part of recovering the lost "crazy feelin'," a spiritual reconnect with ancient societies imagined to have lived in a boundless techno-free world without racism and hatred, imbued only with peace, love, and unity. In 1999 ravers assembled at Ozora, in Hungary, for what was billed as the last total eclipse of the millennium, at the first "Solipse Festival," an ecstasy-filled psychedelic experience centered on the 2^m23^s of totality on August 11.

The Solipse Festival held in Zambia in 2001, organized by a group of German ravers, drew five thousand people for the total eclipse on the June solstice. Raver Gregory Sams described the moment of ecstasy when the moon's first bite silenced the sound system: "Whoops and shouts erupt as the moon takes its first tiny bite from the Sun. . . . Excitement mounts, but it all really explodes at the moment the eclipse goes from 99% to 100%—TOTALITY—and you throw off your silly glasses and stare straight at the black hole that was once just your Sun, and marvel at the electro-magnetic vibrations of its corona. . . . Several couples became engaged, many special batteries were recharged, many spirits restored, many resolutions made—it brought us all together."[11]

In 2002, ten thousand Solipse revelers showed up at the Outback Eclipse in South Australia in search of the techno beat. Despite all the political and cultural conflicts in the world, an event promoter declared, "We will enjoy a natural event together and dance together. . . . It will be like a meeting of the tribes from people all over the world. . . . It should be simple and tribal, a humbling experience that gives you a strong sense of the smallness of each human life."[12] The J, or "techno-shaman," is central to the success of the party, for it is his or her duty to coax guests on the dance floor toward the ecstatic altered state of consciousness leading up to totality, many already having ingested ecstasy to assist them toward the desired communal oneness.

Graham St. John, a sociologist of religion who has written extensively on the spiritual dimensions of neo-paganism and rave culture, sees it as a revitalization movement.[13] Participants stem from a variety of socioeconomic backgrounds, but they unite on activist issues to oppose globalization, logging of forests, uranium mining, junk food, and other ecological concerns. St. John believes the youth culture phenomenon of raves holds deep religious purpose and meaning for its participants, and witnessing the quiescent grandeur of nature's out of the ordinary happenings is part of their spirituality.

The word *eclipse* (*ekleipsis,* from the Greek) implies fear; it means "failure," in the sense of something going wrong. Seeing an eclipse leaves you breathless even if you know what's going on. Like the thrill of a roller-coaster ride, you know you're safe, but the sense of alarm looms within you. Most cognitive scientists think the emotion of fear comes from the desire to find an explanation for any unusual experience. It all goes back to our prehistoric ancestors' need to survive: Is that rustling in the brush that frightens me a wild animal about to attack, or could it merely be a wind gust? Perhaps our belief in transcendent forces came from our basic human need to posit a cause for every effect that would help quell fear of the unknown.

During the 1851 eclipse in Sweden, astronomer François Arago mentioned this aside: "A little girl was watching her flock when the sun began to be darkened. As it gradually lost its light she became more and more distressed, and when at length it disappeared altogether her terror was so great that she began to weep and cry out for help. Her tears were still flowing when the sun sent forth his first ray. Reassured by his light, the child signed herself with the cross, exclaiming, in the *patois* of the province, 'O, beou Souleon!' (oh, beau soleil!)."[14]

On the other hand, philosopher Bertrand Russell once wrote of religion: "I regard it as a disease born of fear and as a source of untold misery to the human race. I cannot, however, deny that it has made some contributions to civilization. It helped in early days to fix the calendar, and it caused Egyptian priests to chronicle eclipses with such care that in time they became able to predict them. These two services, I am prepared to acknowledge, but I do not know of any others."[15] (Incidentally, there isn't a single documented reference to eclipses, much less their being predicted, in the whole of Egyptian history.) Some say religion is an irrational survival of the ingrown habit to find agency. Others think the natural cause argument is a theory proffered by those who are certain science is the only way to acquire truth. Whether the belief is true or not has nothing to do with its origin.[16] So goes the debate between science and religion.

There is an apocalyptic Bible passage connected to the popular term "blood moon" (referring to the red color of the eclipsed disk): "The sun shall be turned into darkness and the moon into blood, before the great and terrible day of the Lord" (Joel 2:31).[17] This passage struck fear into the hearts of many believers in 2014–15, when a spate of books around the event of a lunar tetrad, or four total lunar eclipses in a row covering a period of eighteen months, proclaimed that the Second Coming was at hand.[18]

But this was not the first modern "end of the world" eclipse. On February 4, 1962, a total eclipse over the southern Pacific was accompanied by a rare conjunction of the five visible planets: Saturn, Jupiter, and Mars on one side, and Venus and Mercury on the other side of the sun, all within sixteen degrees of one another. Psychic Jeane Dixon predicted that a child born that day somewhere in the Middle East would unite all warring people in an all-embracing faith that would bring peace to the world by 1999. Meanwhile, some Indian astrologers predicted a global disaster.

In Los Angeles, which experienced only the partial phases, rumors set off a near-panic. The associate director of the Griffith Observatory, Robert Richardson, wrote, "Weeks beforehand we began getting inquiries from people wanting to know, 'What was going to happen on February 4th?' 'What does it mean? When was the last time this happened?' " As the day drew closer, the frequency of phone calls increased and the questions got more detailed: "Will it be dark all day? Are the planets near the sun going to explode? Will they crash together and send the earth off its axis? I'm in a sweat: Can you say something official about it? Will there be big tides? Is it true people are moving inland from the coast? Are we entering a new age? Will Los Angeles be affected?"[19] With the partial eclipse due to happen near sunset, by late afternoon of the fourth (a Sunday), the observatory parking area was filled with eclipse chasers, and traffic leading to the Griffith hilltop site became gridlocked. By 8:30 in the evening, most Angelenos became convinced the earth would survive. In India, however, angry mobs attacked the deceptive astrologers, and a local ruler ordered those who had predicted his demise thrown in jail.

Many contemporary astrologers still hew closely to eastern astrological eclipse wisdom. The Spiritual Science Research Foundation advises followers to avoid consuming food not only during an eclipse but up to twelve hours prior to totality (nine hours for three-quarters partial, six hours for half, and three hours if the sun is one-quarter covered).[20] This is because the light level during eclipses throws the basic subtle components of the spiritual universe, detectable only by those who possess a sixth sense, out of balance, giving rise to the influence of demons, devils, and negative/black energies that sow the seeds of destruction of humanity. These negative energies affect both swallowing and digesting food, not to mention sleeping, sex, and going to the bathroom.

Other contemporary astrological takes on eclipse watching are more upbeat: the sun of life joining with the mysterious quality of the moon during an eclipse indicates change, a time to begin anew. It gives one a chance to review the progress of one's life and "make decisions, rise to the challenge, make an effort, change, mature, take on greater challenges."[21]

When the battery on your smartphone runs out, or the cable service gets shut down, what is there to do? When the lights go out, we lose the constant focus we have on what's directly in front of us. This might be a good time to think about alternatives—actions and choices we don't ordinarily confront because we're too busy with mundane activities. So think of an eclipse as a stop sign, a red light, a brief trip out of the ordinary, or a prompt to rethink how you're looking at things.

"As above, so below"—the idea that what happens in the heavens influences the human condition, would be, I think, the best bumper sticker to characterize worldwide belief in astrology. I'm no proselytizer, but I've written two books that trace the long relationship between astrology and its once sister discipline, my own field of astronomy.[22] The key word here is *influence,* literally a "flowing into." The positions and movements of celestial luminaries don't control us, nor do they predetermine the future. In my opinion, modern popular astrology has more to do with self-help, how we might think about reacting to potential.

Scientifically minded astronomers cast a wary eye on astrology. They see it as a misfit ideology in an age of belief in rational processes. But a 2009 Harris poll showed that 26 percent of all Americans believed in astrology, similar to the findings in a 2005 Gallup poll.[23] A recent survey shows that the percentage of Americans who believe astrology isn't scientific dropped from 62 percent in 2010 to 55 percent in 2012. Americans are less skeptical about astrology than they have been at any time since 1983.[24] (Could it be because the terms astronomy and astrology are often used interchangeably?) I don't think you can lift astrology out of its historical or cultural context and plug it into our contemporary worldview without changing its meaning. My interest in writing about it, like my passion for eclipse lore, stemmed from a desire to explore what it means to those who practice it.

Historian-philosopher Henri Frankfort wrote: "When the sunset is inseparable from the thought of death, then dawn is a sure sign of resurrection."[25] Most of us can scarcely imagine that the two might be linked, except, as they say, "symbolically." But some of us believe all things are connected in an animate universe, that we mortals vibrate sympathetically in response to celestial phenomena. Your mentor, sports counselor, psychologist, or your astrologer, if you have one—those who perhaps know you better than you know yourself—

might say on the day before an imminent eclipse: "Let's talk about your eclipse signs."

Chasing eclipses in cruise ships began in 1972 when eclipse impresario Ted Pedas, an educator-journalist from Farrell, Pennsylvania, organized (along with family associates) a "Voyage to Darkness." Pedas chartered the Greek ship *Olympia* to sail out of New York with a cadre of astronomers, along with 834 passengers on board, to rendezvous with the total eclipse of July 10, 1972, in the North Atlantic. Thanks to an onboard meteorologist, the captain was able to position the ship under a hole in the clouds, allowing the group to view 2^m36^s of totality. Initially, Pedas experienced some resistance from travel agents, who believed it would be difficult to sell clients on undertaking a long summer journey to a usually overcast, frigid place, with the added risk of going blind if the sun did make an appearance. The ship's captain was skeptical too, but for a different reason: "*No one would ever* attend a lecture on a cruise ship."[26]

At the time, I was situated on dry land in Nova Scotia preparing to view my second total eclipse—ironically, in the same place where I'd witnessed my first. Antigonish, Nova Scotia, was the fortunate locale where the March 7, 1970, and July 10, 1972, eclipse tracks crossed, two total eclipses in two years! Pure luck positioned me and a van full of my students under a clear blue spot between a pair of cloud banks. We had spent an exorbitant $15 to secure a spot at a campground on a hill overlooking the provincial highway. Once totality began, we noticed that truck operators barreling along at top speed turned on their lights. When daylight returned two minutes and twenty seconds later, they instinctively switched them off without slowing their speed for the duration. We wondered whether they realized what was happening. Eclipse apathy may be as infectious as eclipse ecstasy.

I didn't become acquainted with eclipse impresario Pedas until the mid-1980s, by which time he had launched several successful astronomy education theme cruises. He invited me to journey as a ship's lecturer to a favorable location to view Halley's Comet, which was due for a close passage by our planet in 1986. Persuaded that the strongest guarantee for eclipse-viewing success lay in the mobility of travel by ship, I had already served as a lecturer on a cruise to the mid-Pacific Columbus Day eclipse in 1977. Margaret Mead, Carl Sagan, and Frank Drake, organizer of SETI, the Search for Extraterrestrial Intelligence, along with an assortment of astronauts and other celebrities, were the main attractions. I was thrilled to have been invited—mostly, I think, because of my Maya expertise. Sailing out of Miami, we made a stop in Yucatán.

By the 1980s, Pedas had secured three ships. I had the pleasure of lecturing on many of Pedas's educational cruises, and I credit him with being among the first to offer an alternative to the typical "cruise, booze, and snooze" leisure ocean voyage.

Pedas always made a supreme effort to reduce the financial burden of professional astronomers of modest means (like me) by paying all expenses, including airfare. He told me the story of how difficult it was to convince Mrs. Koseoglu, the tight-pursed owner of Sun Line, to shoulder these costs: "But don't you realize that these people are *internationally* known lecturers?" Mrs. K. snapped back: "Well I need to save some money. Why don't you just hire *nationally* known lecturers?"

New York Times science writer Walter Sullivan was among those on board the *Canberra* (with two thousand others) for Pedas's second eclipse cruise, to Africa's west coast on June 30, 1973. Also aboard were astronauts Neil Armstrong and Scott Carpenter, sci-fi author Isaac Asimov, astronomer/UFO expert Allen Hynek—and a genuine lunar rock. "For a science freak like myself," Sullivan wrote, "it was 'nerd-heaven.' "[27] He loved the idea of normal ship activities, like bingo and gambling, being overtaken by classes taught by experts who were able to make oceanography, astronomy, meteorology, and African anthropology accessible to interested laypeople.

Despite the advent of super-sized ships, theme cruising is still popular. You can find cruises with the themes "Dancing with the Stars Champion," "Turner Classic Movies," "Cat Lovers," "Mark Twain," and "Lewis and Clark"—and eclipse cruising is alive and well, too. Two eclipse sojourns in 2016 included an expedition to northern Indonesia and a trip sponsored by *Sky and Telescope* magazine. Today, public participation in eclipse watching for the sedentary viewer is widespread, with educational media and Internet live streaming of spectacular events reaching millions of viewers all over the world.

The beginning of the next chapter, which is a bit technical, is intended for those interested in why eclipses happen when and where they do. Later in that chapter, I'll turn to precisely what there is to see under the shadow, the phenomena that attract eclipse chasers of all stripes to situate themselves, whether via stagecoach or cruise ship, within the hundred-mile-wide, pencil-point shadow the moon scratches out on the surface of the earth. But to learn more about the science behind solar eclipses, you need not attend any lectures, on land or sea. The basics are accessible to even the neophyte astronomer.

3

What You See and Why You See It

Beam of the Sun! O thou that seest from afar, what wilt thou be devising? O mother of mine eyes! O star supreme, reft from us in the daytime! Why hast thou perplexed the power of man and the way of wisdom by rushing forth on a darksome track? Art thou bringing a sign of some war, or wasting of produce, or an unspeakably violent snow-storm, or fatal faction, or again, some overflowing of the sea on the plain, or frost to bind the earth, or heat of the south wind streaming with raging rain? Or wilt thou, by deluging the land, cause the race of men to begin anew? I in no wise lament whate'er I shall suffer with the rest!

—Greek poet Pindar, fifth century BCE

What enthralls me most when I witness a total eclipse of the sun is realizing that its precious moments are fleeting. There is too much to see and react to in too little time. Every eclipse starts with a lot of chatter about who sees it first, that teeniest lunar nibble that dents the bright circular disk. It helps if you're apprised in advance of where on the sun's round face to spot it. First contact (the instant when the moon's edge touches the sun) is almost always reported prematurely by naked-eye observers, but moments later the collective shout means we all saw it. Astronomers invariably ask themselves how precisely they predicted it. Deviations can happen because the edge of the lunar disk is ragged with mountain ridges, and there are tiny variations in the moon's orbit caused by the gravitational pull of the other planets.

As the moon slowly begins to munch its way through its round platter, we all begin to hunger for the most savory climactic moments that lie ahead. The time between first and second contact (when the moon just covers the sun) is the longest hour and a quarter wait you'll ever experience. Mylar filters and welders' goggles between eye and light (the smoked glass we used to use when I was a kid has been declared unsafe), we watch the crescent sun imitate the moon as it gradually thins down.

I've never really noticed much dimming in sky brightness up to the three-quarters point between first and second contacts, and I

Phases of a total solar eclipse shown in sequence. (© 2008 Fred Espenak, MrEclipse.com)

wonder whether, had they not been forewarned, people would even have been aware that an eclipse was taking place. Measurements indicate that ten minutes before totality, the light level is about 10 percent of what it would be in broad daylight, but still nearly 50,000 times as bright as a full moonlit sky.[1] Nonetheless, within the ten-minute mark leading up to second contact, several effects begin to be noticeable. As the crescent slims, shadows in the landscape start to sharpen. This is because the normally extended disk of the sun, which casts diffuse shadows, starts to behave more like a point source of light. So shadows of buildings on the ground and the nose on your neighbor's face sharpen and become more stark, and that face takes on an ashen look, as if being viewed through sunglasses.

I recall standing under a tree in bloom during the 1972 Nova Scotia eclipse watching thousands of crescent images of the sun cast on the ground by tiny spaces between the foliage. Squirrels scurried around, seagulls squawked, the air temperature dropped several degrees, the breeze picked up, and the sharply outlined crescents began to shimmer. On November 3, 1994, during one of my eclipse trips off the coast of Argentina, a group of enthusiasts spotted a few dozen bright crescents cast by tiny holes in the canvas cover under which we'd taken temporary refuge from the tropical sun. They instinctively pulled a white-topped table beneath the canvas and continued to follow the show on its surface. You can get the same effect by puncturing a tiny hole in one end of a shoebox, pointing it at the sun, and creating an image on a sheet of white paper taped to the other end, a pinhole camera. Outside the tent, my cohorts amused themselves by discovering they could project sharp shadow outlines of their extended hands with fingers held together, displaying tiny crescent images emanating from the spaces between them onto the floor. One woman cast a hundred crescents through tiny holes in her straw hat—magical!

As the rapidly dimming light begins to change color, turning a ruddy brown or metallic gray, next come the shadow bands, so named when they were first reported in 1820. During the Columbus Day eclipse of October 12, 1977, several passengers and I on board *Sitmar's Fairwind* laid out a king-size bedsheet on deck. About two minutes before (and after) totality, we spotted the faint alternating light and dark bands about six inches apart moving erratically across the stark white background, reminiscent of ripples reflected off the bottom of a pool on a sunny day, especially when the water has been slightly disturbed. On March 29, 2006, a group of German eclipse chasers shot some of the best photographs of shadow bands. They were able to determine that the waves lined up parallel to the crescent sun. When the air just above ground level is moving slightly, the effect is reminiscent of "heat waves," the same phenomenon that causes stars to twinkle.[2] Eclipse watchers rarely notice these delicate bands because they are preoccupied anticipating the main event, which looms more moments ahead. We measured the shadow bands on our bedsheet by placing a yardstick on its surface. Once we began to view the pattern, we adjusted the ruler so that it lay perpendicular to the direction of movement of the bands. A guest used his stopwatch to time how long it took for a band to travel the length of one of the sticks, three miles an hour. (How fast they move depends on wind speed.)

For the total eclipse of March 29, 2006, we drove all day west from Cairo, a police escort behind and ahead of our bus and an armed guard on board. We ate and slept in a roadside hotel and arose at 1:30 a.m., in order to arrive at the coastal town of Sallum, near the Libyan border, where the shadow was scheduled to pass two hours later. We stationed ourselves along with three thousand others in one among a vast complex of tents, their floors covered wall to wall with thick decorative rugs so our feet would never touch the sand. Women in burkas served tea, and camels grumbled in the background amid the chatter during the long wait between first and second contacts. About half an hour before second contact, we heard a chopper approaching from the east. It set down on a pad not more than a hundred yards from us, close to where a podium had been set up on a stage. Egyptian president Hosni Mubarak climbed down from the cockpit, and his guards assisted him to the platform. He made a short welcoming speech to the guests and praised the power of nature for treating his domain to one of her most magnificent spectacles. He was cutting it close, as moments later the real show began in earnest.

About ten minutes before second contact, we noticed the moon's shadow approaching in the distant landscape. It resembles a dark, silent storm on the western horizon. Two minutes before totality, the light level plummeted and darkness started to envelop us. (The effect reminds me of what I like to do when the light of a chilly dawn awakens me out of a deep sleep—pull the warm comforter over my head and descend back into the darkness.) The final approach of the lunar shadow always thrills me beyond expectation. Here comes that anticipated silent wall of blackness at screaming speed. Some say it's like heaven descending to earth.

The eclipse chaser Rebecca Joslin described her feelings on witnessing the eclipse of August 30, 1905, in Spain: "We hardly had time to draw a breath, when suddenly we were enveloped by a palpable presence, inky black, and clammy cold, that held us paralyzed and breathless in its grasp, then shook us loose, and leaped over the city and above the bay, and with ever and ever increasing swiftness and incredible speed swept over the Mediterranean and disappeared in the eastern horizon. Shivering from its icy embrace and seized with a superstitious terror, we gasped, 'What was that?' . . . *THAT* was the Shadow of the Moon!"[3]

The temptation to look directly at the solar crescent in the minute before it begins to disintegrate is overwhelming. Don't do it! If you start too soon, you'll get a lasting afterimage that will rob you of seeing much during totality. In 2006, ten seconds before second contact, the last remnants of solar light skirted the jagged edge of the moon as sunbeams leaked out through the valleys between peaks on the lunar perimeter. The crescent was now nearly a complete circle, and it began to break up into tiny rounded spots, what are called Baily's beads, after the English astronomer Francis Baily, who popularized them in his description of the 1836 eclipse in Scotland. Instead of viewing the anticipated thread of light as the rim of the sun disappeared, he reported seeing instead "a row of lucid points, like a string of beads irregular in size and distance from each other, suddenly formed round that part of the circumference of the moon that was about to enter on the sun's disc."[4] The solar necklace is visible only for an instant. The last pearl to vanish looks like a jewel set into a ring of light—the "Diamond Ring," as New York City eclipse watchers would dub it in 1925.[5] The label stuck.

Then comes second contact. The sun as we know it is gone; protective glasses come off as day becomes night! "A vast palpable presence seems overwhelming the world," the wife of astronomer David Todd, Mabel Loomis Todd, wrote. "All the world might well be dead and cold and turned to ashes . . . as the world holds its breath. Then out upon the darkness, gruesome but sublime, flashes the glory of the incomprehensible corona, a silvery, soft, unearthly light, with radiant streamers, stretching at times millions of uncomprehended miles into space, while the rosy flaming protuberances skirt the black rim of the moon in ethereal splendour."[6]

The sun's plasma envelope, or corona, is the main spectacle during totality's most precious moments, which range from less than a second up to 7^m29^s, the time between second and third contact (the instant when the moon is tangent on the inner side, its edge just kissing the place on the sun's circular disk just opposite of where it entered). The corona is partially visible up to several seconds before totality, but now it takes over the celestial lighting duties. Blending with the ruddy illumination at the far horizon in the partial eclipse zone, the strange skylight cast a livid hue on the faces of those around me.

Observers have difficulty expressing the color of the corona. To me it looks like a bright greenish-white halo surrounding the black disk, but others have characterized it as pearly, silver gossamer, and a color that white lacks.[7] Rebecca Joslin was a Smith College student who traveled to Spain to view the August 30, 1905, eclipse, then returned to war-torn Europe for the August 21, 1914, event. She lost out to clouds on both attempts before catching the big one in New York City in 1925. She wrote of the corona: "Gold it was, and not the pearly, silvery crown we had been told to expect."[8] I think much of the confusion about the corona's color has to do with the composition of coronal light. Though, like normal sunlight, it is comprised mostly of a continuous spectrum made up of all the colors of the rainbow, the spectrum of the corona includes a small component composed of discrete bright lines, especially a pair of concentrated emissions in its green and red portions due to highly ionized iron atoms produced in the extremely hot environment of the corona. Likewise, energy-efficient compact fluorescent lamps (CFLs) and other fluorescent light are made up of such selective wavelength emissions, and they cast distinctly different color tones on whatever they illuminate.[9]

At the peak of totality, coronal streamers up to half a dozen sun diameters long radiate outward from the fully eclipsed disk. They curve and delicately interlace as if some cosmic artist had rendered them in brushstrokes. Usually at sunspot maximum, every eleven years, the streamers are arranged more or less evenly along the perimeter of the disk, while at minimum, as we witnessed them in 2006, they fan out along the sun's equator, appearing chopped off and bristly, like a crew cut, at its poles.

The corona is essentially the sun's outer atmosphere. It is now known to be hotter than the visible surface of the sun (the photosphere), with temperatures above 900,000 degrees Fahrenheit, compared with some 5,800 degrees Fahrenheit on the photosphere. Astronomers still puzzle over how the energy involved in raising the temperature gets transferred to the corona.[10] Between the photosphere and the corona lies the chromosphere. This consists of gases that interact with the sun's magnetic field, licking out like flames into the corona along the black lunar rim. They were first characterized (in 1831) as "red flames" and likened to "a mighty flame bursting through the roof of a house and blown by a strong wind."[11]

Most spectators spend the precious few moments of totality *documenting* (my eclipse pet peeve) instead of *seeing*. During the August 11, 1999, eclipse, which was total from southern England, France, Germany, and the Balkans out across the Black Sea through Turkey, we floated calmly under clear skies awaiting totality. As soon as it grew dark, both sea and sky were lit up (unnecessarily) by hundreds of flashbulbs, as everyone sought to record totality with the latest photographic equipment, from telescopes equipped with fancy Canon and Nikon cameras down to the ubiquitous, then-popular disposable cameras. Based on what I could hear within thirty feet of where I was standing, I estimated 20,000 shutter clicks during $2^m 20^s$ of totality (there were 750 in attendance). I kept repeating over the loudspeaker, "Totality!—totality! . . . look up, look up . . . take a moment to look at it!" But the documenters remained as riveted to their cameras and telescopes as many of us to our smartphones during public events today. The few who did look away from their apparati saw the stars come out. A small group of enthusiasts insisted on remaining in an indoor lounge, where they monitored the eclipse on laptops, their preferred medium for experiencing reality.

All too soon third contact happens, and it's almost always followed by applause (accompanied, in the case of the Egyptian eclipse, by the sound of Mubarak's chopper starting up). The diamond ring reappears toward the opposite side of the disk. It widens once again into Baily's beads, which shift position to the opposite side now that the disk of the moon has migrated across the sun. Time reverses its course and the shadow bands return, this time gliding along the ground opposite their course prior to totality. The somber lunar shadow swiftly takes leave, racing to the east to meet the distant horizon; the temperature slowly rises as the wind picks up again. Most observers on my trips, their senses beleaguered by all they've taken in, appear too numb to pay attention to the replay in reverse of each phenomenon they witnessed moments before.

All is quiet. Then, with increasing enthusiasm, as we emerge from the shadow we begin to share experiences of what we saw and felt. "Amazing corona; I counted five streamers, maybe six. Did you see the long one at the four o'clock position? Must've stuck out half a dozen disks. Was that Vega off to the west? I think it was one of the planets. Did you catch those beads?" Less concerned with detail, a young woman once hastened over to me, tears streaming down her cheeks: "You may have told us all about the science of eclipses, but for me it was all a miracle!" An elderly woman confessed, "I became so excited when I heard you shout totality, but I forgot to take off my protective glasses"—long pause—"but I still enjoyed it!"

The time between third contact (when the moon is tangent to the disk on the opposite side) and fourth (when the moon entirely departs the solar disk and the partial phase of the eclipse officially ends) is usually spent making new friends and reconnecting with old ones. Umbraphiles already begin to plan where in the world they intend to venture to secure their spot under the next shadow. Tales are shared by those who were clouded out on previous eclipses: one man traveled all the way to Siberia for fifty seconds of totality on August 1, 2008, only to see the cloud-bound sky darken briefly. I've also met a number of people who got rained out in the Florida Panhandle during the eclipse of 1970—an ancestor of the great American eclipse of 2024.

To understand why eclipses happen, hold a pea at arm's length, and you'll see it just about covers the full moon. It also fits perfectly over the sun. Everybody knows the sun is actually much larger than the moon, four hundred times larger, but it also happens to be four hundred times farther away. Because both the moon's orbit around the earth and the earth's path around the sun are elliptical, the sizes of the two disks seen from the earth's surface vary slightly.[12] The near perfect fit means that when the moon passes between the earth and the sun, it either totally blocks out the sun's light with little room to spare, or, if it happens to be farther away than usual, it doesn't quite cover the sun. These are the conditions for a solar eclipse. An eclipse of the moon takes place when the earth comes between the sun and the moon and the moon passes into the earth's shadow.

Elsewhere in the universe, eclipse circumstances aren't so tidy. Of the 163 moons that orbit six planets in the solar system, ours is the only one that just happens to be a near perfect fit to the sun's disk. Observers on Mars can witness tiny satellite Phobos eclipsing the sun. It was photographed doing so by the Mars rover on August 17, 2013.[13]

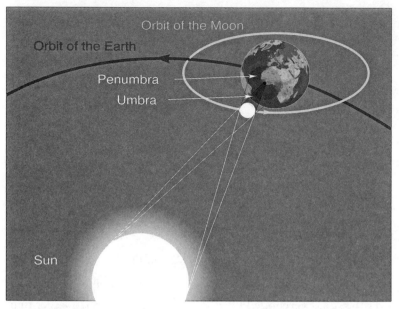

A solar eclipse happens when the moon comes between the earth and the sun. As the blunt point of the darkest part of the shadow (the umbra) moves over the rotating earth, viewers within it witness a total solar eclipse, while those in the lighter portion (the penumbra) see a partial eclipse. When the earth intervenes between the moon and the sun, and the moon enters the earth's shadow, a lunar eclipse occurs. (Wikimedia Commons)

The eclipse lasted a little over twenty seconds, but the irregular shape of the mini-moon covered less than half the solar disk.

With sixty-three moons to gaze at, hypothetical eclipse chasers situated at the top of Jupiter's perpetual cloud deck could experience lots of eclipses, but not one of the moons perfectly frames the sun, which, at Jupiter's distance, is only one-fifth as wide (and one-twenty-fifth as bright) as seen from earth. Tiny, sausage-shaped Amalthea makes the best fit, provided it's positioned the long way. Rounder

An annular eclipse of the sun as seen from Mars. (Photo courtesy of NASA/
JPL-Caltech/Malin Space Science Systems/Texas A&M University)

Callisto appears twice the size of the sun in the Jovian sky. The other
so-called Galilean (larger) satellites are way too oversized as well. For
Saturnian observers, Janus comes close, but its irregular shape is 20
percent too large, and Titan lives up to its name, covering more than
twenty-five times the area of the sun's disk, which, nearly a billion
miles away from Saturn, measures only one-tenth the size of what
earthbound observers see. Most satellites are not only irregularly
shaped but also too tiny to cover the sun when viewed from their parent planets. Like the porridge-loving bears in the tale of Goldilocks,
of all the views that heaven allows, only the sun and our solo satellite
are positioned just right to create a perfect-fit total eclipse. Coincidence, or did God design it that way? Mathematician David Orrell
thinks this is strong evidence we are living in a virtual reality.[14]

Because the sun and moon disks are almost exactly the same size in the sky, when the moon passes in front of the sun, it casts a shadow that comes to a point close to the surface of the earth. It can be as wide as 167 miles when the moon is closest. If the moon happens to be relatively far away, however, the tip of the shadow comes to a focus before it hits the ground, the sun isn't entirely covered, and you get an annular eclipse, so named because a thin ring, or annulus, remains visible around the eclipsing lunar disk. In the long run this phenomenon won't last, because the tides cause the moon's distance from earth to increase an average of about one foot per century. At that rate, in a little over a billion years from now the lunar disk will fall short of fully covering the sun—no more daytime darkness!

Even under optimum conditions, not many people get to see a total eclipse. Though the eclipse path travels across the world, it's so narrow that the odds your home base happens to lie in the line of fire are very small. Of course, you could opt to stay put and see a partial eclipse, but that's a bit like being in the stadium the day *before* the Super Bowl.

Or you can wait for the next one. About 240 solar eclipses happen somewhere in the world in a century, and about a quarter of them are total. Residents of Jerusalem have been patient for 884 years; they will need to wait another 225 years for totality (on the other hand, there were three total eclipses in that city in the fourth century BCE). If you really want to improve your chances, pack your bags and pursue the most convenient means of getting under the shadow, provided you know where it's headed.

When and how often eclipses happen depend on how the sun and moon move across the sky, the moon throughout the course of its monthly phase cycle, and the sun over its seasonal journey along the ecliptic through the constellations of the zodiac.

If the plane of moon's orbit about the earth coincided with the sun's path, a lunar eclipse would occur at every full moon, when earthbound observers are positioned precisely between the sun and the moon. At this time the moon would fall within the shadow of the earth, which is nearly triple the width of the moon's face. Solar eclipses would occur two weeks later at new moon, when the moon lies between us and the sun. However, the moon's orbit is tilted five degrees relative to the earth's orbit around the sun. As a result, eclipses can happen only when a new or full moon lies close to one of the two intersection points, or *nodes,* of the two orbits. This occurs periodically, but because the moon and the sun (or the moon and the earth's shadow, if it's a lunar eclipse) are disks rather than points, they can overlap if positioned to one side or the other of a given node. The zone within which solar eclipses can happen is called the solar ecliptic limit.

Foreknowledge is power. Think of the advantage of knowing in advance, the more precisely the better, when natural events will take place. Meteorologists have the ability to forewarn, if not pinpoint exactly, where tornadoes and hurricanes will occur. Geologists have made progress, but still struggle in their attempts to narrow the timeband for forecasting earthquakes along geologic faults. Thanks to the pristine regularity of many celestial phenomena, astronomers have a decided advantage when it comes to the business of prediction. Based on previous similar occurrences, they can calculate when and where most events will happen. Our modern scientific goals in foretelling the behavior of nature focus on saving lives. As you can see from this chapter's epigraph, the same was true of our ancient ancestors, who were more directly concerned with anticipating eclipses as a means of acquiring omens that foretold the luck of kings and queens, the conduct of war, and the outcome of plagues and harvests.

I can't prove it, but I strongly suspect that long before recorded history, people must have noticed that lunar eclipses happen when there's a full moon, while solar eclipses occur during a new moon. Even though they take place with approximately equal frequency, patterns of recurrence of lunar eclipses would have been more easily recognizable because when they do occur, the entire half of the world that faces the moon gets to witness them (actually more than half, because the earth rotates more than one-eighth of a turn during the course of a lunar eclipse). If you pay attention to what's going on in the sky you're apt to spot a lunar eclipse, at least in the partial phase, about every two and a half years; on the other hand, the average interval to randomly witness a total solar eclipse is about four hundred years (much less for partial eclipses).

That eclipses repeat in cycles extending over long periods of time was one of the most significant discoveries made by our literate ancestors. Finding complex patterns that emerge from eclipse timings necessarily would have required persistent sightings, diligent recordkeeping, and a good bit of individual genius. How could they do it? What patterns might our forebearers have detected? Basically, when whole multiples of the relevant lunar and solar time cycles fit together, eclipses must repeat. Suppose, for example, that the sun, positioned exactly at one of the nodes, gets eclipsed by a new moon. Now, the time it takes the moon to return to a given node is 27.21222 days. The ancients called it a *draconic* month, after the dragon believed to devour the sun during an eclipse. The month measured by the lunar phases, the *synodic* month, which is controlled by the sun, is 29.53059 days long. So, by the time the moon is new again, it will already have passed the node. Eclipses recur only when these two cycles become synchronized. For example, if the draconic and synodic periods were whole numbers, say twenty-seven and thirty days, respectively, then an eclipse would take place every 270 days. That's because nine moons (nine times thirty days) would equal ten passages of the new moon by the node (ten times twenty-seven days).[15] In reality, there is no time period that fulfills the desired conditions for generating eclipse cycles perfectly, but some long intervals do come close. Any of these periods could have been recognized, though some cycles might have been more easily detectable than others. The question is: which cycles *did* our ancestors recognize and record?

To address that question, let me offer a simple exercise, free of geometry and complex computations, that shows how you can tap into the power of forecasting eclipses the way our ancient ancestors did. Begin by looking up all the lunar eclipses, total or partial, visible from your area over, say, the past ten years.[16] Suppose you had witnessed every one of them. Make a list of the dates. Next, determine the number of days between each eclipse in the sequence. (This isn't so tedious if you make use of one of the many Julian Day converters available online. All you need to do is enter the date and read out the corresponding Julian Day, basically a running long number; for example, the lunar eclipse of September 28, 2015, occurred on Julian Day 2457294, and the previous one happened on Julian Day 2457117. If you subtract one from the other, you'll find that the eclipses happened 177 days apart.) Next, make a list of all the intervals and start looking for patterns. My ten-year sequence of intervals, which includes eleven eclipses, reads like this (I live in upstate New York at latitude 42°49′N, longitude 75°32′W): 502, 178, 177, 177, 679, 178, 531, 680, 176, 178, 177. As I looked over these numbers, I immediately noticed a majority of intervals of 177±1 days (or six lunar synodic months) between successive eclipses.[17] Long in use as seasonal dividers in ancient calendars, this six-month period is known as a *semester,* from the Latin, meaning six months. (The calendar in our college curriculum grew out of this two-part division of the seasonal year.) As I looked more closely at the numbers, I realized that one of the larger intervals, 531, is exactly three semesters long, and that the others, 502, and 679–680 are, respectively, two semesters plus a five-month interval of 148±1 days (let's call it a short semester), and three semesters plus a short semester.[18] Hindsight knowledge of the geometry of orbits helps account for the short semester. First of all, recall that eclipses happen in an "eclipse vulnerability" zone centered on the nodes, known as the ecliptic limit, rather than only at a point (the node itself). It takes the sun more than a month (approximately 34 days) to traverse the ecliptic limit. Second, the line connecting the nodes regresses; that is, it moves slowly opposite the direction of motion of the sun and moon. Suppose the sun is near one node on January 1 of a given year and that a lunar eclipse takes place on that date. The next time a lunar eclipse can occur, the sun will be at or near the opposite node. But because the line of nodes regresses, completing a circuit in 18.6 years, in one year it will have moved 1/18.6 × 360°, or 19°.4 westward; therefore, the sun will arrive back at the first node about 19 days short of a year, or 346⅔ days later, known as an eclipse year. It will reach the opposite node 173⅓ days (an eclipse half-year) after January 1, or June 23. Around this time the second lunar eclipse of the calendar year would occur. Because there must be a full moon, this eclipse will most likely take place six months, or 177 (or 173 + 4) days after New Year's Day, on June 27 or, less likely, five months (148 days) later, on May 29. A third eclipse may happen about December 12, 346⅔ days after January 1. Consequently, as many as three lunar eclipses can occur in a given calendar year.

When I extended my ten-year observational base to fifty years, once again I found a majority of visible lunar eclipses one semester apart and a number spaced several semesters (especially six and seven), plus one short semester, apart. I made it easy for myself by including partial eclipses. Also, because I used a computed list of modern eclipses, I didn't need to worry about cloudy weather preventing me from seeing an eclipse. Given the actual observations of generations of ancient skywatchers to add to my own, I could certainly acquire a significant data base. But did I miss something? Do eclipses *always* happen at semester intervals, with occasional short semesters sprinkled in? And is there a pattern that reveals where short semesters fall in the sequence? To answer these questions, let's turn to the evidence in the historical record.

Babylonian astronomers were not as concerned with orbits and geometry as our later Greek ancestors. They did eclipse arithmetic instead, basically what I've been doing with my list of archival lunar eclipses. By the sixth century BCE, Babylonian astronomers' diaries were filled with records of historical eclipses viewed by their predecessors, much lengthier lists than the one I laid out in my simple example. By the fifth century BCE it had already become clear to the record keepers that eclipses of both the sun and the moon recurred at semester or, more rarely, short semester intervals. They noticed, as I was on the verge of discovering when I lengthened my data base from ten to fifty years, that either 41 (6 + 6 + 6 + 6 + 6 + 6 + 5) or 47 (6 + 6 + 6 + 6 + 6 + 6 + 6 + 5) months following a total eclipse, another nearly identical total (or almost total) eclipse took place. When they lengthened their observational time base, they were able to refine their conclusions: they realized that following a total of two series of the first kind coupled with three of the second kind, an almost identical sequence of eclipses recurred. This long cycle of 2 × 41 + 3 × 47, or 223 lunar synodic months, or 6,585⅓ days (18 years, 11 days, 8 hours to be exact), ran for up to fifteen centuries.

Later astronomers named this cycle the *saros*, after the Greek word meaning repetition. One saros also encompasses 242 passages of the new (or full) moon by a node. (As I mentioned earlier, any closely matching whole [or half] multiple of each of the two periods will yield an eclipse cycle; for example, 38 node passages = 35 lunar synodic months, a 1,033.5-day cycle; and 439.5 node passages = 405 lunar synodic months, an 11,960-day cycle. The latter cycle was noted in ancient documents written by Maya astronomers, who preferred it, for reasons I'll discuss later, to the saros, though there is evidence that they knew about the saros as well.)

The saros is easy to recognize because it is a *seasonal* cycle; in other words, the eclipses happen at the same time of the year, backing up only eleven days every eighteen years. Because of the extra one-third of a day, or eight hours, a succeeding eclipse in the saros series occurs 120 degrees (⅓ of an earth rotation period) west of the original longitude. Over three saroses, or 54.09 years, an eclipse in the series returns roughly to the same place on earth. Additionally, the period of variation in the size of the moon disk, and hence the length of its shadow, varies over a nine-year cycle, so that after 2 × 9 or 18 years, similar durations of totality also recur—yet another reason why the saros cycle became so conspicuous.

To summarize, we can think of eclipses occurring in families. If we regard the first eclipse we observe in a saros as the parent, then the offspring would take place at the same time of year, and each would resemble its predecessor. For example, take the first of the contemporary pair of great American eclipses, August 21, 2017. Its saros parent must have lived eighteen years minus about a week and a half earlier. The records show that there was indeed a total solar eclipse on August 11, 1999, the last one of the 1900s (though it was erroneously billed as the last of the millennium). Its path touched down off the coast of

Nova Scotia, tracked across the Atlantic, and moved on through central Europe. The midpoint, where a maximum totality of 2^m23^s occurred, was situated in the Balkans (that's the eclipse I watched from a ship off the Bulgarian coast in the Black Sea). The moon's shadow departed earth in the Indian Ocean. Totality at the midpoint of the 2017 eclipse (2^m40^s) will take place in the southern great plains approximately one-third of an earth rotation (or 120 degrees west) of its predecessor. To find a closer relative—a similar eclipse that repeats at the same time of year *and* in the same general area of the world—you need to back up one triple saros, or 54.09 years, to the "great-grandparent" of 2017. That would be the total eclipse that took place on July 20, 1963. Its path of totality was fairly parallel to the anticipated 2017 track, except that it ran about twenty degrees farther north. Totality lasted 1^m40^s. Maine was the only state where the eclipse was visible (total for a little over a minute), so we can officially call it a mainland American eclipse. Maine resident and writer Stephen King featured it in the plots of two of his novels. The 1963 eclipse resurfaced in a 2009 episode of the 1960s nostalgia series *Mad Men*. The great-grandparent of the April 8, 2024, eclipse (totality 4^m28^s) happened on March 7, 1970 (total for 3^m28^s). Its track landed to the east of where the 2024 event will be positioned; it ran from the Florida panhandle, where it was a total washout, up to the Carolina and Virginia coasts. (This was the eclipse a group of students and I videotaped when the shadow got to Nova Scotia.)

Saroses last an average of thirteen centuries and include seventy to eighty-five solar eclipse members in all. Up to forty-two saroses run simultaneously, and astronomers refer to them by number. For example, the family belonging to Saros 136 is famous for having produced a sequence of long durations of totality in the contemporary era. It includes the July 11, 1991, event (at 6^m53^s), which touched

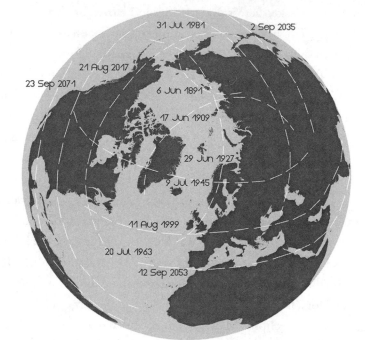

An eclipse in the next saros happens about 120 degrees west of its parent. Move forward three saroses and the great-grandchild of the original eclipse occurs at nearly the same longitude. Here we see three families of projected eclipse paths in Saros 145. One family includes eclipses occurring in the years 1891, 1945, 1999, and 2053. A second family includes eclipses occurring in 1909, 1963, 2017, and 2071. A third family includes eclipses occurring in 1927, 1981, and 2035. Eclipse predictions by Fred Espenak (NASA's GSFC). (Basemap data © OpenStreetMap contributors [CC BY-SA].)

Hawaii and central Mexico, in addition to the 7^m04^s north African eclipse of June 30, 1973, and that of June 20, 1955. The longest in the series at 7^m08^s, it was visible in southwest Asia. First born in that family was a 5 percent partial eclipse that took place on June 14, 1360. The first total (for one second) in Saros 136 happened on

November 22, 1612. Now that we've crossed into the third millennium, this series has slowly begun to wane. Next comes the July 22, 2009, eclipse, which will offer up only 6^{m}39^s of darkness to residents of India and China. The last in the series of total eclipses in the 136 family will take place on May 13, 2496, and the very last (a 10 percent partial) will occur on July 30, 2622—ending an impressive 1,262-year run.[19]

Though every eclipse belongs to a saros family, like physical appearances of succeeding human descendants, facial resemblances change slowly with time because the relevant cycles don't fit together perfectly. Durations of totality lengthen or shrink slightly through time. For example, check out the eclipse paths spaced one triple saros apart in Saros 145, to which the August 21, 2017, eclipse belongs.

Go back to the 1909 and 1963 family ancestors of 2017, and you'll find that the paths have gradually progressed southward as the durations of totality get longer. If you look ahead to the great-grandchild of 2017, the eclipse of September 23, 2071, you'll notice that the path of totality will have departed the U.S. mainland and migrated across the border into Mexico. Also, as the moon's closest distance from the earth increases, its apparent size shrinks, and so does the duration of totality (see Table 1). The opposite is true for portions of Saros 139, the April 8, 2024, eclipse family. The tracks trend more gradually from south to north, and move slightly to the west; then, beginning with the 2078 event, they regress and go southward again.

With sufficient knowledge of the science behind eclipses, you can now relive what other people have seen through human history. And don't worry if you don't understand all the details associated with the saros cycle, as long as you're aware that eclipse recurrence patterns

Table 1 Triple saros series members in close proximity to the 2017 (Saros 145) and 2024 (Saros 139) eclipses

Saros/member no.	Date	Duration	Location
145 (16)	June 17, 1909[a]	0^{m}24^s	Russia, Canada, Greenland
145 (19)	July 20, 1963[a]	1^{m}40^s	Northeast Asia, Canada, Central America, Maine
145 (22)	August 21, 2017[a]	2^{m}40^s	North America
145 (25)	September 23, 2071[a]	3^{m}11^s	Mexico, northern South America
139 (24)	February 3, 1916[b]	2^{m}36^s	Northern South America
139 (27)	March 7, 1970[b]	3^{m}28^s	North America, Central America, northwestern South America
139 (30)	April 8, 2024[b]	4^{m}28^s	North America

Source: "Solar Saros 145," available at en.wikipedia.org/wiki/Solar_Saros_145, and "Solar Saros 139," available at en.wikipedia.org/wiki/Solar_Saros_139. The tables link to a map of each eclipse.
[a] This is a young cycle, beginning on January 4, 1639, with a 1 percent partial; it will peak with a 7^{m}12^s event on June 25, 2522, and end with a 6 percent partial on April 17, 3009.
[b] This saros began on May 17, 1501 (9 percent partial), and will end on July 3, 2763 (6 percent partial), peaking in totality at 7^{m}29^s with the eclipse of July 16, 2186.

do exist. When I boarded a cruise ship to view the 1999 Black Sea eclipse, a passenger, evidently out cruising for reasons other than my own and apparently unacquainted with the fact that eclipses recur, seemed convinced I was wasting my time: "Why are you here now? Don't you know the eclipse happened eight years ago?"

CONTACT TWO
Eclipses in the Ancient World

• • •

4

Eclipse Computer Stonehenge

Every age gets the Stonehenge it deserves or desires.

—*Jacquetta Hawkes, 1961*

As you enter Salisbury Plain along the A303 from London, you can't miss Stonehenge. Stark and brooding, the multi-ton megaliths that make up its circle have dominated the landscape for five millennia. Accompanied by colleagues from British Heritage, I had the opportunity to retrace the stones' journeys to this now famous site. We walked the two miles from the Avon River, where the giant stone slabs were disembarked after being hauled from various quarries in the surrounding mountains between 20 and 130 miles from their current resting place. It's a curved uphill climb that gradually straightens out as you approach the site. As we got closer, we could see the tallest of the standing stones, looming thirty feet into the sky. I tried to imagine processions of people passing several abreast along the wide access, perhaps on the way to celebrate a ceremony in the enclosure.

As we reached the end of the causeway, we passed through what once was an imposing pair of standing stones, the first megaliths erected at Stonehenge, set in place about 2600 BCE. Only the easternmost, the Heel Stone, remains. Sixteen feet of it lie above ground level, and it weighs about thirty-five tons. Slightly tilted, the Heel

Stone stands like a lone sentinel more than seventy-five yards outside the closely gathered rings of worked stones at the center of the ruin. Next, via another pair of stone portals, we entered the largest break in the ditch-and-bank enclosure that surrounds the main stone complex. Was this portal intended to keep bad guys out, or perhaps good thoughts in? Since the nineteenth century, archaeologists who have excavated Stonehenge have debated the possibilities.

Continuing our walk into Stonehenge's inner sanctum (a fence prohibits entry today), we next encountered the so-called Aubrey holes, fifty-six evenly spaced chalk- and rubble-filled pits between 2.5 and 6 feet wide and 2 to 3.5 feet deep, arranged in a circle just within the periphery of the enclosure. Named after the sixteenth-century antiquarian who first uncovered them, the holes were dug about 2900 BCE. They remain so precisely laid out that the circle that guided their placement surely must have been ruled with a rope and stake anchored at the monument's center.

We could barely make out the four large Station Stones arranged in a perfect rectangle on the perimeter of the Aubrey circle, its short axis aligned with the main avenue. Radiocarbon dating indicates they originate from the same time as the Heel Stone. A double ring of thirty and twenty-nine holes (the so-called Y and Z holes) lie within. Once we broke through the ring of Aubrey holes, we needed only about three dozen steps (twenty-five yards) to enter the next major circle at Stonehenge, an impressive ring forty-five yards in diameter consisting of thirty Sarsen stones.[1] These car-size sandstone slabs stand sixteen feet high and weigh twenty-five tons apiece. Constructed about 2450 BCE, they are capped by horizontal slabs, or lintels, all the way around. Archaeologists think the builders dragged these stones via sleds from Overton Down, about twenty miles north of their current location.

Passing farther inward, we next encountered a double circle of eighty-five bluestones from the Preseli Mountains, lugged across approximately 240 miles. Weighing in at about four tons apiece, these basalt blocks were set up vertically about six feet apart in a pair of concentric circles consisting of thirty-eight members per ring.

But it was not until we took the last few steps into the monument's center that we encountered Stonehenge's most celebrated megaliths, the five trilithons that make up a giant horseshoe-shaped arrangement. Constructed out of the longest of the bluestones, the trilithons (so called because each is made of three stones: two vertical supports and a capstone) weigh in at close to fifty tons apiece. The vertical supports are positioned so close together that you can't wriggle your body in the vertical space between them. The horseshoe, twelve yards wide, has its open axis pointing straight toward the northeast-aligned causeway.

The innovative architect who conceived the plan and the influential chief who amassed the labor force to construct it evidently had particular tastes when it came to the diverse choices of building materials, and they were quite fond of circles. If you visit the ruins that dot Salisbury Plain, you'll discover a shared passion for circularity. The timber Woodhenge and Durrington Walls nearby are both circular in form. The latter, a village enclosed in a circular ditch-and-bank structure, even has a similar accessway. (A semicircular arrangement of fifteen-foot-long megaliths dated to 2500 BCE has recently been excavated at Durrington Walls.)[2] Of course, Stonehenge replaces impermanent wood with everlasting stone from the Preseli Mountains.

Stonehenge wasn't built in a day. Diverse cultures constructed, deconstructed, and reconstructed this site over two millennia. The earliest phases were laid out by the Windmill Hill people, who crossed the English Channel a thousand years before Stonehenge was built, bringing the first sheep and goats to the British Isles. They lived in sod houses and buried their dead in round mounds. Next came the Beaker People, semi-sedentary bronze tool users from central Europe. They were well organized into little chiefdoms. The even more hierarchically organized Wessex people entered the picture by 2100 BCE. They raised cattle and traded extensively with people on the mainland. They were the likely innovators of the trilithon horseshoe.

Stonehenge is a good example of multifunctional architecture. Over the ages it has been interpreted as a place of pilgrimage, a ritual center dedicated to divinatory practice and worship of the dead, an economic center, a place of fortified habitations, a celestial temple, an observatory, and a computer for tracking eclipses.

No one has ever doubted that Stonehenge, at least at some stage, had something to do with skywatching. That explanation has been around at least since the early eighteenth century, when clergyman-antiquarian William Stukeley pointed out that the axis of the trilithon horseshoe and the direction of the causeway aligned with sunrise on the June solstice, the longest day of the year. Visitors still gather in the circle to witness sunrise over the Heel Stone on the first day of summer.

The eclipse prediction theory of Stonehenge entered the stage of the cosmic drama in the early 1960s, when British astronomer Gerald Hawkins concluded that the monument was designed as an observatory to mark significant positions of both the sun and the moon as they periodically shift along the horizon.

Along with the day-night cycle and the phases of the moon, the sun's annual path in the sky is one of the most universally recognized cycles in the natural environment. Like a swinging pendulum, sunrise and sunset positions change with the seasons, shifting along the horizon from south to north in winter and spring. They reach a stopping point, the summer solstice (from the Latin for "sun standstill") on

Ancient skywatchers at Stonehenge view the June solstice sunrise over the Heel Stone. (Drawing by Allie Pushlar)

June 20; then they reverse direction and pass southward along the horizon through summer and fall, as the sun heads for its southern-most standstill on the first day of winter, December 20. I've pointed out many instances where this annual back-and-forth solar move-ment served as a convenient environmental calendar for marking im-portant dates of planting and harvesting, and related religious rituals and festivals, in various cultures of the world.[3]

The importance of predicting and following seasonal change among prehistoric people cannot be overestimated. In societies de-void of technology, time was an expression of activity. It was *lived* rather than kept track of on a mechanical device and treated as an external entity—the way we think of it. (Native Americans once called our September and October full moons Harvest Moon and Hunter's Moon, respectively.) Getting a direct sense of internalized time from watching the sky is difficult because we no longer need practical astronomy in our daily lives. We pay little attention to the heavens, except maybe as an occasional diversion. We might catch a sunset, glimpse the man in the moon, and perhaps notice the evening star. But who would think of using such casual observations to set up appointments or schedule social or business activities? The artificial clocks that mete out our daily activities leave us with an underappreciated view of our ancestors' dependence of timekeeping on phenomena that happened in the world around them.

With a site map of Stonehenge, Gerald Hawkins calculated several alignments and found that they matched up with directions he obtained by drawing lines between prominent megaliths. Using the center point of the site as a backsight and the Heel Stone as a foresight, he discovered that the alignment between the two coincided precisely with sunrise on June 20, as expected. He found the same result for an alignment taken between Station Stones 93 and 94, while the direction from 91 to 92 pointed to midwinter (December 21) sunset.

Then came a surprise: out of a total of twenty-four, fourteen of Hawkins's alignments matched the standstill positions of the moon. Like the sun, the moon also rises and sets cyclically. It takes approximately 18⅔ years to complete the cycle, and the lunar standstills stretch out over a wider portion of the horizon than those of the sun.[4] The lunar alignments between Station Stones 91 and 94 and between 93 and 92 ran perpendicular to the solar standstills. Other alignments taken through the trilithon and Sarsen archways also matched up with the solar and lunar standstills. Hawkins discovered that opposing solar and lunar standstills ninety degrees apart are possible only at the

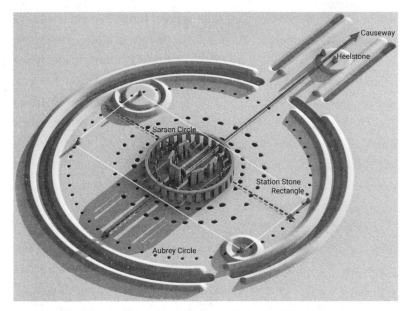

Plan of Stonehenge showing its principal parts and key sun and moon
alignments. (Wikimedia Commons)

latitude of Stonehenge. If you travel farther north or south and lay out
stones to mark these astronomical directions, you get a parallelogram
instead of a rectangle. Did the builders deliberately seek out a place
where they could create a "magic rectangle," or was it a coincidence?

A young scientist growing up in an age of problem solving via the
novel medium of electronic calculation, Hawkins looked at Stonehenge and saw a computer. Here were fifty-six evenly spaced Aubrey
holes arranged in a perfect circle. Could they have been part of a
Neolithic analog computer device used to predict eclipses? If so, how
did it work? Hawkins found that if you placed stones in various holes
at the proper spacing and shifted them annually, you could track a
56-year eclipse prediction cycle made up of 18, 19, and 19-year periods,

which average out to the 18⅔–year cycle of lunar standstills already present in the alignments. For example, take three white stones, a, b, c, and set them at Aubrey holes 56, 38, and 19, counting clockwise from the center of the causeway. Then take three black stones, x, y, z, and set them at holes 47, 28, and 10. By shifting each stone one position every time the sun rises over the Heel Stone as seen from the center of the site (that is, once a year), you can predict significant lunar events. For example, the full moon nearest winter solstice would rise over the Heel Stone whenever any stone is positioned at hole 56 (every nine years). And when it does, there will usually be an eclipse of the sun or moon within fifteen days of winter solstice in that year (that is, during the month containing the full moon). Hawkins's cycle keeps in step with eclipses because the time spacing he chose, 9, 9, 10, 9, 9, and 10, measures out intervals of 18, 19, and 19 years, which average out to 18⅔ years, for the black or the white stones. This is the same as the time it takes the moon's rising or setting to make one full oscillation from northerly to southerly standstill and back again, a cycle of 18.61 years, which, Hawkins contended, also governs eclipse frequency.

By tabulating dates of eclipses and positions of moonrise, Hawkins demonstrated that such a prehistoric stone computer could work perfectly for about three hundred years, after which its predictive capacity would fall behind one or two days. But ancient Stonehengers could easily have corrected that by intercalating—that is, by advancing all the stones one position. Bolstering Hawkins's theory, noted astronomer Sir Fred Hoyle wrote an accompanying piece in the British journal *Nature* enhancing the theory with an alternative (slightly more complicated) computer model. Later, Hawkins published his controversial research papers and followed with a best-selling paperback, *Stonehenge Decoded.*[5] He tells the story of how he pioneered the use of the newly designed mainframe computer to study astronomically aligned ancient architecture.

A widely viewed television production, *The Mystery of Stonehenge,* shows Hawkins standing in front of a massive mainframe computer uttering, "Generations of archaeologists have puzzled over the meaning of Stonehenge. . . . Well, *I* have the answer." He concludes: "The odds that I'm right in my explanation are millions to one in my favor."[6] *Stonehenge Decoded* includes a chapter simply entitled "The Machine," in which Hawkins refers to how he applied the latest technology to decode Stonehenge: "This work was made possible by the donation of approximately one minute of time on the Smithsonian-Harvard electronic computer"—a very impressive statement at the time.[7]

It makes sense to imagine prehistoric Bronze Age sun and moon watchers. But were they watching eclipses too? Their academic turf violated, archaeologists and prehistorians weighed in with criticism: just because a modern astronomer finds celestial alignments and a mechanism using the Aubrey holes to compute eclipses doesn't mean that such knowledge actually was known to the builders of Stonehenge. Most of the excavated chalk-covered Aubrey holes contain cremated human bones, which hardly supports the Hawkins-Hoyle eclipse computer theory. What about the cultural motive? Why would a semi-sedentary tribal people need to keep track of rarely observed long-term sky events? Were they a race of curious prehistoric Einsteins motivated by a desire to explore the mysteries of the universe *for its own sake,* like modern scientists? And why isn't there any other evidence of astronomical practice among them? Many critics felt that the ancient wise men Hawkins envisaged were fabricated cardboard ancestors of himself, projections of his own contemporary worldview in an age of computer-assisted, precise, quantitative science. Jacquetta

Hawkes, who wrote a scathing review of Hawkins's work, concluded: "Every age has the Stonehenge it deserves—or desires."

As chief archaeologist in charge of Stonehenge excavations for more than two decades, Richard Atkinson had every right to dive into the fray. He also criticized Hawkins's book in an essay, cleverly titled "Moonshine on Stonehenge." Atkinson complained that Hawkins drew his conclusions from measurements made on a touristic map, quite unsuitable for acquiring his claimed precision, and that he never set foot on Stonehenge with measuring equipment. As for the odds that fourteen alignments pointed to the moon, Atkinson noted that he calculated them to be much less—using only a pencil and a piece of paper. He summed up Hawkins's work as tendentious, arrogant, slipshod, and unconvincing. In cultural astronomy circles, Atkinson's piece is still regarded as the nightmare of all reviews.[8]

Notwithstanding the harsh criticism, Hawkins pioneered the study of ancient astronomy in the archaeological record. His work led to the use of the term "astroarchaeology," or the study of astronomy in the unwritten record and, oddly enough, to my ultimate involvement in that field. I first encountered Gerry Hawkins in the late 1950s, when he was a young professor (without any apparent interest in Stonehenge), and I a freshman at Boston University hungering for astronomy courses. Hawkins refused to allow me a place in his radio astronomy course because he felt I lacked the appropriate math background; he was probably correct. Following his brief moments of academic celebrity, Hawkins eventually left Boston in 1969 and accepted a deanship at Dickinson College, fully exiting the halls of academia two years later.

In the 1990s, while I was conducting field research on the lines of Nazca, the giant ground drawings on the coastal desert of southern Peru, I met Hawkins for lunch. He had written a piece on his study of Nazca alignments that I had included in one of my edited books. We

mused about the pinnacles and pitfalls of engagement in interdisciplinary research, the thrill of attacking a set of problems outside your
normal sphere of inquiry, and the price you pay when you tread in-
cautiously on alien disciplinary turf.

As I look back, I think it was Hawkins's brash confidence that ignited the turf war between astronomers and archaeologists over the
function and meaning of prehistoric architecture. He told readers, for
example, "If I can see any alignments, general relationships or use of
the various parts of Stonehenge, then these facts were also known to
the builders."[9]

Having followed the Gerald Hawkins saga, I have learned to tread
more carefully. You need to make a sincere effort to understand, and
appreciate, the theories and methods employed by your colleagues for
acquiring answers to the *kinds of questions they ask*. Dealing with the
prehistoric mind is far more complex than studying atoms, rocks, and
galaxies.

Though Stonehenge was built in prehistoric times, there is some
recent circumstantial evidence of early time reckoning that predates
it. In 2009 a group of archaeologists excavated an alignment of pits in
northeast Scotland dated from the eighth millennium BCE. They
were found to point to the December solstice sunrise position over a
prominent cleft in the hills along the local horizon. The number of
pits, twelve, is the same as the nearest whole number of lunar synodic
(phase) months in a seasonal year, and they are so arranged that they
could have been used as a counting device to compensate for the seasonal drift associated with the annual succession of months.[10] While
this is a long way from an airtight case of prehistoric eclipse predicting, it can be regarded as a beginning.

Recent excavations at Stonehenge have bolstered the astronomical hypothesis. In 2013 archaeologists unearthed manmade ditches

along the processional route, confirming that the site was deliberately constructed over an ice-age landform aligned along the solstice axis. The land seems to have pointed out the connection to the sky. An active period of exploration begun in the 1980s continues to reveal other henge-like monuments over a much wider surrounding landscape, including the super-henge at Durrington Walls. During its early phases, Stonehenge may have served as a burial ground as well.[11] The cremated bones of sixty-three individuals were found within the bluestone circle. Also, recently excavated tombs from around the site contain remains of elite individuals from foreign lands.[12]

There isn't space here to review all the research and discussion of the meaning of Stonehenge, except to remark that, as we cross the threshold of time into the sixth millennium of its existence, the eclipse prediction theory remains on the table, and it still arouses controversy.[13] Having followed the literature for more than forty years, my own opinion is that the association between Stonehenge's architecture and the sky may have been more closely allied with theater than exact science. Too often we neglect the power of religion in framing responses to natural phenomena that lead to the environment we construct to celebrate their occurrence. Religion is far more than a simple byproduct of human cognition without function. It serves as a way of coping with the mysterious forces in the universe that move the sun and the moon and create startling phenomena like eclipses.

I think Stonehenge was built at least in part to celebrate the entry of the sun god—or goddess—into the circular sanctuary. It was designed to chart the deity's seasonal back-and-forth course along the horizon, as well as that of its celestial complement—the one with the slivered, silvery variable countenance, which migrated even farther to the northern and southern climes. In southern Britain, the June solstice would have been the best time to anticipate a bountiful crop; on that day the sun would then follow its highest noonday track as it slanted across the sky, officially initiating the peak of the growing season.

I wonder whether, by chasing eclipses in stone, my fellow astronomers might have looked past a more fundamental aspect of moon watching that builds sensibly on the Stonehenge alignments with the sun's horizon extremes. If you follow the way it courses through the seasons, you'll find that, in addition to the bright light it provides, the full moon also mimics the sun in mirror image. It passes high in the sky around the time of the winter solstice and arcs low in the sky during summer. You always see the full moon rising directly in front of you just as the sun disappears behind you. For a short time around sunset, both the sun and the moon appear opposite one another on the horizon. It's easy to notice if you pay attention to the environment.

Seeing is believing, and it is little wonder that in cultures all over the world the sun and moon make up the two halves of a cosmic duality. On the night of the full moon, the lunar deity takes over the duties of lighting up the world. Just as daytime's great illuminator vanishes from view, the lunar disk rises, glowing ever brighter as it ascends. This is the day of the month that bears light from dawn to dusk, then back to dawn again. I think winter solstice full moonrise at Stonehenge is even more impressive than summer solstice sunrise. That moon rises within ten degrees of the Heel Stone, and it occurs on the same day the sun sets opposite it in the trilithon archway. The count of the Y and Z holes located just outside the Sarsen circle, thirty and twenty-nine in number, could have been used alternatively to mark out, again cleverly in the round, the days of the 29½-day lunar month. When the count of these holes, commenced at the opening, returned back to the entryway to the inner circle, another full moon would be scheduled to appear.

But did the builders of Stonehenge go further? Did they recognize the lunar standstills and the 18⅔-year cycle of the moon? If so, did they discover a way to use this period to predict eclipses, perhaps even devise a rudimentary computer for generating such predictions? Given the evidence, I can't be sure. But if more information is unearthed and the answer eventually turns out to be yes, I won't be surprised at the scientific capabilities of prehistoric watchers of the skies.

The biggest question of all is: where does skywatching fit into the worldview of the people who built Stonehenge? Today's historians, archaeologists, and cultural astronomers have a more integrated understanding of Britain's megalithic architecture, one that brings together land and sky. Many of Bronze Age Britain's monuments are oriented to other monuments and landscape features as well as to the sun and moon. Archaeologist Julian Thomas wrote of "the momentary coincidence of chalk from the earth, the descending sun, the dead in their barrow and the surrounding forest." Such architecture serves as an expression of the "unity of earth and sky, life and death, past and present, all being referenced to more and more emphasis onto particular spaces and places," the idea being to create a theatrical environment for the ritual performances that once took place there.[14]

Stonehenge remains as famous and enigmatic as ever. I've visited several replicas of it—and even designed one. In 2002 Jonathan Rothberg, a pioneer in genetic sequencing technology, asked me to help create his "Circle of Life," a half-scale Stonehenge replica he planned to erect in his expansive backyard at Sachem's Head in coastal Connecticut. He'd already arranged to import seven hundred tons of Norwegian granite from across the Atlantic for the job. When I arrived at the site, surveying equipment in tow, I asked how close a replica Rothberg desired: did he want me to lay out alignments to the lunar extremes so he could predict eclipses? "No," replied the biologist-turned-DNA decoder; "I want the alignments to point to where the sun rises and sets on the birthdays of my wife and kids," he said.

"And yours?" I inquired of his birthday.

"No, you can leave that out."

When I pointed out that there was an excellent quarry in North Branford, a few miles down the road from where he lived, he gave the same opinionated response the tribal chief of Stonehenge might have offered his head architect: "I prefer the stone from over there."

The Circle of Life was completed in 2004. It still stands, despite a lawsuit filed by neighbors to remove it on the grounds that it was an eyesore. When I attended the gala affair on the summer solstice of 2004 to celebrate the Circle of Life's dedication, I told Rothberg that while the site had certainly been built to last, and the alignments I calculated would remain valid for at least three thousand years, his backyard would almost certainly be inundated within a century by the rising waters of the Atlantic. He shrugged and upturned his palms.

Why is it, I wonder, that everyone wants to lay claim to the great works of the past? Maybe it's because it boosts your social status in the present. The claimed connection between Stonehenge and the Druids was forged in the popular imagination during the seventeenth century. Annual visits by Druids to the original monument have been dated to the late nineteenth century, when romantic inventors of tradition dressed in robes began to show up to watch the summer solstice sun rise over the Heel Stone. Fraternities not unlike the Masonic Order, they became more spiritually oriented social activists, especially on environmental issues, by the turn of the century.[15]

The astronomical theory of Stonehenge has long appealed to popular New Age interests in alien civilizations with a scientific understanding of the world far beyond our modern capabilities. During the 1970s, New Agers, like the ravers of the 1990s, were holding live concerts attended by tens of thousands and camping out by the side of the highway leading across Salisbury Plain to the ruins. Loud music, nudity, and vandalism (including spray painting megaliths) accompanied the month-long gatherings surrounding the June solstice date, which led to the establishment of a camping exclusion zone and a fence denying access to the stone circle. Most Druids today see themselves as the descendants of pagans who come to Britain's most sacred ancient place to exercise their religious rights. In 2008, a confrontational group of Druids protested the removal to a museum display of the bones of their ancestors recently excavated at the site by archaeologists. A more moderate group claimed that, as descendants of the original builders, they should have been consulted. The archaeologists countered that, unlike many Native American tribes, their claim to genetic links with excavated ancestors' bones is undocumented. Druids also appear, especially at summer solstice, at replicas, such as the one near Goldendale Observatory in Washington State, one of eighty large "clonehenges" around the world.[16] An especially large contingent was present in 1979 when the most recent United States mainland eclipse, total in the northwest, took place. Dressed in long white robes and hoods, they beat their drums and chanted in unison to bring back the disappearing sun deity.

Finally, speaking of eclipse coincidences, the August 21, 2017, eclipse path is due to pass over Alliance, Nebraska, home of Carhenge, an oft-visited avant-garde work of architecture just off U.S. Highway 385, in which discarded automobiles, positioned vertically, take the place of standing stones. A popular website advertises a celebration on the big day.

5

Babylonian Decryptions

When Tiamat opened her mouth to devour him
He drove in the evil wind, in order that she should
not be able to close her lips.

The raging winds filled her belly;
Her belly became distended and she opened wide her mouth.
He shot off an arrow, and it tore her interior.
It cut through her inward parts, it split her heart.
When he had subdued her, he destroyed her life.

He split her open like a mussel into two parts;
Half of her set in place and formed the sky therewith as a roof.
He fixed the crossbar and posted guards;
He commanded them not to let her waters escape.

—Enuma Elish, *tenth century BCE*

Enuma Elish, a tenth-century BCE Babylonian story of creation
written on seven tablets, recounts the slaying of Tiamat, female goddess of the waters of the Persian Gulf, by Marduk, the protector god
of the city of Babylon. The story follows the once orderly evolution,
personified in the lives of the gods, of the landscape between the
Tigris and Euphrates Rivers, in what is now Iraq. There, silt slowly
builds up to form land in the peaceful boundary where sweet and salt

waters mix. The established order is suddenly overturned by violence. Heavy winter rains create a watery chaos, causing the rivers to overflow their banks. But in the spring, Marduk, god of storms and wind, unleashes his power to dry the land; he parts the waters and creates the sky out of Tiamat's murdered carcass, as order is restored. Both scenarios are experienced in every season of the year.

The script of this cosmic drama emerges out of a state committed to establishing rule over its rivals by military force. *Enuma Elish* weaves two different versions of creation, one placid, the other violent, together in a single myth. The narrative represents both society and nature. Once the gods, under Marduk's leadership, put things in place in the terrestrial domain, they took up residence in the sky. For example, Ishtar's (Venus's) home was a place that radiated sexuality and fecundity. In contrast, the dwelling place of Nergal (Mars) was a violent domicile that generated malevolence. Shamash, god of justice, ruled the daytime sky, while Enzu, lord of wisdom, reigned at night. Keeping track of the positions and movements of these temperamental deities was the job of the court astronomer, a specialist in astral religion trained to track both the denizens of heaven and the signs, or omens, they emanated. Despite mythical motivations, the astronomical side of astrological prophecy, knowing where heaven's denizens are located, how they move, and, most importantly, when they coalesce, was hard science. Cracking the code of eclipse prediction lay at the top of the list of dynastic necessities.

Compared with measurements of alignments among the megaliths of prehistoric Stonehenge, evidence of eclipse watching in *historical* cultures (those with writing) is much more reliable. Still, there are difficulties. When we try to pinpoint early eclipses in archival records, we need to correlate inscriptions based on different time schemes with our own calendar. Using the available information, astronomers can make calculations to determine the conditions that best fit what they find in the written record.

The medium of written expression is usually fabricated out of the most convenient, lasting material available. The Egyptians wrote on papyrus, the pithy stem of a plant that grows abundantly along the banks of the Nile. The Inca of Peru kept records on knotted strings made of cotton, the staple crop in the valleys that run down the Andes to the coast. Maya codices are made of lime-coated bark of ficus trees that flourish in the Yucatán Peninsula. In the land between the Tigris and Euphrates, where the earliest written records date to the fourth millennium BCE, the medium was mud.

Before the invention of irrigation, springtime floods plagued Sumeria. The Zagros Mountains in northwest Iran emptied torrential rain and silt into the Tigris and Euphrates to commence the planting season. In valleys abutting the rivers, the moisture-laden, thick clay soil dried up in the intense early summer heat, creating mud cracks. The smooth surfaces of the thick plates of hardened clay proved smooth and durable, and especially inviting to one who might pick up a firm reed or stick, press it into the soft medium, then drag it over the surface, using it as a stylus to sketch an animal in the herd he was tending, or his hut, or perhaps a species of fish he pulled out of a nearby stream. When the stylus got stuck in the mud and snapped in the subsequent dragging motion, the artist-cum-budding writer realized that pressing without dragging was a more efficient way of making lasting impressions on the viscous clay. Sumerian epigraphers agree that something like this scenario led to the invention of cuneiform writing, after the wedge-shaped (Latin: *cuneus*) stylus used to impress characters on clay tablets. Pictographs made up of linked wedge-shaped markings gradually evolved into abstract representations of phonetic and later alphabetic symbols.

The earliest cuneiform tablets date from late in the third millennium BCE. Unlike the surviving Maya documents, which deal with calendars and religious rituals, the corpus of Sumerian cuneiform writing consisted of bills of lading, inventories of sacks of barley, herds of sheep, jars of oil, and items related to business, trade, and the economy. Thousands of broken tablets discarded by traders ended up as rubble fill in the construction of houses and buildings in cities along trade routes.

Once you get a closer look at the profound corpus of Sumerian writing, it becomes obvious that all these texts are just notes about market transactions. A few hundred tablets, like the Venus Tablet of Ammisaduqa, named after the seventeenth-century BCE Babylonian king, contain astronomical information.[1] This document contains a list of dates, dutifully compiled by a court astronomer over a period of twenty-one years, of the celestial appearances and disappearances of Ishtar, goddess of love. Keeping track of Ishtar, nearly always seen lying close to the horizon at dawn and dusk, must have lain close to the top of the professional skywatcher's agenda. Ishtar's choreographed movements were thought to harbor potential answers to a most important set of questions about what the afterlife is like: Where do the sun, moon, and stars go when they pass over the western horizon? What subterranean journey do they undertake between their disappearance in the west and their reappearance in the east? What happens in that unseen netherworld, the place where our departed souls are destined to venture? And what do these wanderings tell us about the future of the world? Questions about divinatory matters captivated the astronomer-astrologer's attention.

A story told around the fire has it that Ishtar, beckoned by the sun god, Shamash, removes her veils and descends via the seven gates of heaven (the planetary orbs that surround the world) into the afterworld. There she has an affair with the solar deity. When she fails to return, her faithful messenger, Ninshubur, incites a general clamor in the heavens. Enlil, god of air, and Nanna, the lunar deity, reject his plea, but Enki, the water god, comes to his aid. Out of the dirt beneath his fingernails, Enki creates two sexless creatures. He sends them into the netherworld, where they discover Ishtar's corpse. They sprinkle the secret food of life over her remains. She awakens, fixes herself, and wastes no time making a getaway, retrieving her garments as she passes through the seven gates in reverse order. But the goddess of love takes with her the omens that hold the key to our fate.

The oldest surviving copy of this record of Ishtar's travels, the Venus tablet, is labeled K160 in the British Museum's collection, and is among thousands of cuneiform documents (the Epic of Gilgamesh among them) once housed in the library of Ashurbanipal, last king of the Neo-Assyrian empire in the seventh century BCE. Those of us who have studied the Venus text marvel at its extraordinary detail: a formulaic repetition of the names of Venus, dates, appearance and disappearance intervals, and, usually in the last sentence, the omens. Here is one example: "If on the 25th day of the ninth month the Queen of heaven disappears in the east, remaining absent in the sky two months four days, on the 29th day of month eleven, Ishtar appears in the west, the harvest of the land will be successful."[2] By analyzing the described course of Venus, astronomers have dated the original document from which K160 was copied to sometime between 1702 and 1550 BCE. Despite the absence of telescopes, computers, and indeed anything resembling what we would call technology, the Venus tablet is accurate to better than a single day in a century, thanks to repeated, careful observations made with the naked eye.

Keeping track of eclipses was even more important than following the love goddess, for such happenings veiled the solar god Shamash himself; however, given their rarity, the task of decoding hidden patterns of eclipses would prove far more challenging for them than tracking Ishtar, and it wouldn't happen until two thousand years later.

Those who chase eclipses via the archive agree that the earliest definitive written record of a solar eclipse in the Middle East appears on a clay tablet found in 1948 in the ruins of the city of Ugarit in Syria. The words on tablet *KTU* 1.78 read: "On the . . . day of the new moon in [the month] *hiyaru* the Sun went down, its gatekeeper was Ršp." The flip side reveals the omen: "Two livers were examined: danger."[3] (It was customary for a professional hepatoscopist to divine by reading the markings on the livers of sacrificed sheep, the way one might read your palm or tea leaves, to extract what the event portended.)

Looking at the astronomical side of the statement on *KTU* 1.78, scholars identified the gatekeeper with what we call the planet Mars and *hiyaru* as the month that spanned the second half of April and the first half of May. From other documents, the reconstructed Ugaritic calendar and its place in relation to the Julian (Western) calendar point to a date somewhere between 1250 and 1175 BCE. The only eclipse that fits all the conditions, including Mars's presence in the sky, is one that took place on May 3, 1375 BCE. Enter the "different time scheme" problem: other eclipse historians claim that earlier investigations confuse the Ugaritic month name *hiyaru* with the Babylonian *ajjaru* and point to the eclipse of March 5, 1223 BCE.[4] Dates notwithstanding, *KTU* 1.78 looks more like an eclipse *record* than a prediction.

Long before recorded history, people must have noticed that lunar eclipses happen only when there is a full moon, while solar eclipses occur during a new moon. Also, lunar eclipse recurrence patterns would have been easier to detect because, on average, you see far more of them at any given location than solar eclipses. All science begins with pattern recognition. Finding complex patterns that emerge from eclipse timings necessarily would have required persistent sightings, diligent recordkeeping, and a good bit of individual genius. That eclipses repeat in cycles extending over long periods of time was one of the most significant discoveries made by our literate ancestors from the Middle East. Remember that eclipses happen at semester intervals, with occasional short semesters sprinkled in. The pattern discovered by the Babylonians that gives the location of short semesters in the sequence is known as the *saros,* which consists of a sequence of 41 (6 + 6 + 6 + 6 + 6 + 6 + 5) and 47 (6 + 6 + 6 + 6 + 6 + 6 + 6 + 5) months, specifically two series of the first kind coupled with three of the second kind. This long series of 223 lunar synodic months, or 6,585⅓ days (18 years, 11 days, and 8 hours, to be precise), lasts up to fifteen centuries, until the sun departs the end of the node and the series breaks down.

The most solid evidence for the detection of the saros eclipse cycle by Babylonian astronomers comes from a fragment of text named the "Saros-Canon." Also housed in the British Museum, it covers dates from 373 to 277 BCE and was probably written down after 280 BCE. The Saros-Canon consists of a vertical list of years (32, 33, 34, and so on) and months (X, IV, X, and so on), with abbreviated names of the ruling kings attached. Table 2 shows a partial transcription of its contents. The symbol "dir" indicates that an extra (thirteenth) month was inserted to fit the lunar synodic month more precisely into the year. Note that the months sometimes increase by six (for example, X + 6 = IV [first column, line 2], IV + 6 = X [second column, line 3]), and at other times by five (X + 5 [one month inserted] = II, so marked in the first column, line 4). Each full column contains 38 lines, making a total of 223 months, or one saros. Furthermore, lines separating the months break the sequence down into the 41- and 47-month series (two of the first kind and three of the second) that make up a saros.[5] If you compare this table with modern computed lists of lunar eclipses, you'll find that a total eclipse occurs at the middle of each series.

The Saros-Canon proves unequivocally how the realization of a fundamental eclipse series first came to light based on a knowledge of patterns of eclipses obtained from historical records of observed

The earliest known record of an eclipse, one that took place in Babylon on April 15, 136 BCE, appears on a cuneiform tablet.
(© The Trustees of the British Museum)

Table 2 Partial transcription of
Saros-Canon

	X	
32	IV	
dir	X	
33	II	5 m
	VIII	
34	II	
dir	VIII	
35	I	
	VII	
36	I	
	VII	
	XII	5 m
37	VI	
dir	XII	
38	V	
	XI	
39	V	
	XI	

eclipses. It also reveals a process early Middle East astronomers developed—let's call it a theory—to project both backward and forward in time to pinpoint lunar eclipses. Studies of other Babylonian tablets since the reading of the Saros-Canon document have revealed that by the second century BCE, astronomers were able to determine the saros with extraordinary accuracy.[6]

Now that we know for sure that Babylonian astronomers were motivated to predict eclipses for religious reasons, omen gathering in the service of the state, and that they had successfully arrived at a formula known as the saros for pinpointing lunar eclipses, what about solar eclipses? They could have been most recognizable halfway through the two series mentioned above, or 20½ and 23½ months after a total lunar eclipse. Unfortunately, no document similar to the

Saros-Canon for *solar* eclipses has been discovered to date. This is not surprising, given that these events are rarer and, consequently, not so easily foretold. What I learned from my simple exercise in eclipse arithmetic is that duration is all that stands in the way of progress. I think common sense would have informed any persistent ancient astronomer that both kinds of eclipses, solar and lunar, play by the same rules. Still, we have no direct proof Babylonian astronomers could predict solar eclipses precisely, though astrological warnings for solar eclipses do appear in the literature.

Like the work of today's neurosurgeons, the labors of Babylonian astronomers were a matter of life and death. I can only imagine the stress levels of my ancient counterpart as he went about his demanding task. Fragments of texts culled from astronomical diaries offer a clue that betrays the frustration of one loyal subject:

> The king has given me the order: Watch and tell me whatever occurs! So I am reporting to the king whatever seems to be propitious and well-portending [and] beneficial for the king, my lord [to know]. . . . Should the king ask, "Is there anything about that sign?" [I answer] "Since it has set, there is nothing. . . ." "Should the lord of kings say, "Why [did] the first day of the month [pass without] your writing me either favorable or unfavorable [omens]?" [I answer], "Scholarship cannot be discussed in the market place!" Would that the lord of kings might summon me into his presence on a day of his choosing so that I could tell my definite opinion to the king my lord."[7]

Before moving on to the methods and data of the Greek inheritors of early Middle East eclipse prediction, let me share a diary entry

I came across that captures one desperate diviner's poetic plea to the celestial deity through the animal he is about to sacrifice so that he might verily see into the future of the state:

> O Great ones, gods of the night,
> O bright one Gibil, O warrior Irra,
> O bow [star] and yoke [star],
> O Pleiades, Orion and the dragon,
> O Ursa Major, gout star, and the bison.
> Stand by, and then,
> In the divination which I am making,
> In the lamb which I am offering
> Put truth for me.[8]

6

Greek Science

If someone had said he had performed his ablutions in vain because the sun did not go into eclipse, he would be ridiculous. Solar
eclipses are not what washing is for.

—*Aristotle, fourth century BCE*

Those who chase eclipses in the written record venture over historical landscapes made up of poetic and allegorical contours. Their
sky is often clouded over by undocumented references and misleading
statements. Take the case of Theoclymenus's prophecy in the eighth-
century BCE *Odyssey*.[1] Penelope's suitors are just about to sit down
for a noontime banquet when the goddess Athena confounds
their minds, making them laugh uncontrollably. Suddenly, the guests
notice that their food is spattered with blood. Theoclymenus rises
and makes a speech foretelling the death of the men (by Penelope's
lost husband and the story's eponymous protagonist) and their
banishment to Hades. He ends with the words: "The sun has
been obliterated from the sky, and an unlucky darkness invades the
world."[2] This statement has always struck me as a casual reference.
Fifth-century BCE writer Heraclitus and first-century CE writer
Plutarch called it an "allegorical eclipse," used by the poet for effect.
But I suppose it could have been the vestigial memory of an eclipse

that was total on the island of Leucas, possible site of Odysseus's home base of Ithaca.[3]

Early twentieth-century astronomers developed a penchant for computationally crisscrossing the Mediterranean in search of actual eclipses that fit with the projected date of the storytelling of Odysseus's meanderings. One candidate they turned up was a spectacular total eclipse of the sun that happened in the Ionian Sea on the afternoon of April 16, 1178 BCE—spectacular because all five naked-eye planets appeared strung out beyond the solar corona. Later astronomers used other astronomical references in the *Odyssey* to set constraints on a possible date; for example, there was a new moon before the massacre of the suitors when Venus was high in the sky, and the sailors navigated by the constellation Boötes (with the Great Bear on their left) to get to the island: again 1178 won the day.

There is no concrete evidence that the Greeks had any foreknowledge of astronomical events before writing was developed (about 700 BCE). But things would change by the time Athens developed, and ultimately gifted us, in the fourth century BCE, with the unique way of knowing the universe that we call *science.*

When in the beginning, as their account runs, the universe was being formed, both heaven and earth were indistinguishable in appearance, since their elements were intermingled: then, when their bodies separated from one another, the universe took on in all its parts the ordered form in which it is now seen; the air set up a continual motion, and the fiery element in it gathered into the highest regions, since anything of such a nature moves upward by reason of its lightness (and it is for this reason that the sun and the multitude of other stars became involved in the universal whirl); while all that was mud-like and thick and contained an admixture of moisture sank because of its weight into one place; and as this continually turned about upon itself and became compressed, out of the wet it formed the sea, and out of what was firmer, the land, which was like potter's clay and entirely soft. But as the sun's fire shone upon the land, it first of all became firm, and then, since its surface was in a ferment because of the warmth, portions of the wet swelled up in masses in many places, and in these pustules covered with delicate membranes made their appearance. Such a phenomenon can be seen even yet in swamps and marshy places whenever, the ground having become cold, the air suddenly and without any gradual changes becomes intensely warm.[4]

Roman historian Diodorus's (first-century BCE) hand-me-down story of the classical Greek version of the creation of the world sounds nothing like Babylon's *Enuma Elish.* For example, notice the way the writer uses the sort of scientific language we employ today and ends by referring to what *can be seen* in the world we experience directly. Diodorus's worldview mentions not a single deity or religious rite. It offers instead a rational explanation free of superstition, to account for why things happen in the heavens. Indeed, eclipses are not for washing.

Is the scientific method handed down to us by the Greeks the unique key to explaining how nature *really* works? Is it a method that might be known to any intelligent alien from a faraway galaxy? Or is it purely a product of Western culture, one of myriad ways of accounting for how the world around us behaves? The debate rages on in the halls of contemporary academia, and I have taken up its issues elsewhere.[5] We do know this: the Greeks were different from other cultures of the past in the way they thought about the heavens.

Western science is, in part, a product of what historian of science Derek Price called the "Greek miracle," a way of understanding the natural world based on the dominance of number, derived from the Babylonians and transferred, via logic, to abstract geometrical imagery.[6] This innovation was accompanied by an intense focus on motion in the heavens and an abiding faith that a rational, god-free, underlying basis could be devised to explain it.

Why these circumstances took place in the circum-Aegean area in the first few centuries BCE is also widely debated. The Greek democratic attitude certainly helped condition them to explore the exotic mathematics they inherited from the Middle East, to openly debate different theories of motion, and to tinker with models that would precisely duplicate the orbits they believed gave rise to the motions they perceived in the sky. But chance also played a pivotal role in the coming together of so many of the approaches and skills that produced the core of our scientific approach to the real world. This peculiarity, this fundamental difference between the forebears of our civilization and all others, argued Price, was "surely a spectacular accident of history."[7]

You might not think of searching through a political science text if you wanted to understand what causes eclipses, but it is in Aristotle's *Politics* that our concept of orbiting celestial bodies, like the sun and the moon, takes root. Aristotle theorizes on how to construct the ideal democratic city, the polis. His very first word in the text is one that we associate today with astronomy: "*Revolving* round the marketplace and the city-centre [agora], people of this class [the shopkeepers, mechanics, and day laborers] generally find it easy to attend the sessions of the popular assembly—unlike the farmers who, scattered through the country-side, neither meet so often nor feel so much the need for society of this sort" (emphasis mine).[8]

The agora—that dynamic pivot of motion where all citizens assemble—is the geometrical focal point of the city of Athens (today we'd call it the commons or the town green). Greek democracy situated everything in the public domain, and all matters were subject to open debate and argumentation in the agora. When it came to decision making on civic matters, the priest-king in the hierarchy of power was replaced by the citizens arranged in perfect equilibrium. Theirs was a balanced system. Running the city came to be more a secular than a religious activity. Extending the civic discourse to a dialogue on the cosmos, the earth would become the agora, and the sun, moon, and planets would be likened to the various social groups that occupied their appropriate orbs, or spheres of action, about it. Such a breakthrough in imagining the design of the ordered universe came about as an application of everyday, real-life experience.

Aristotle's philosophy was developed during a time of radical economic and social transformation in the Greek city-state. In the Greek world, nothing dominated the agora—the public power point where each citizen could take his place and speak freely. This idea stood in stark contrast to the hierarchical domination of a king over successive lower classes in the more or less pyramidal social model that preceded urban Athens. Likewise, the organization of the sun, moon, and planets in geometrically measured orbs that go round and round like clockwork was a long way from the imagined universes of other cultures we'll explore—universes worth probing in their own right, even if they are not the particular scientific universe bequeathed the modern West as a gift from the Greeks.

The Greek word *geometry*, from *geo* ("earth") and *metron* ("measure"), reflected the Greeks' love of "measuring the earth." The abstract geometry many learn in high school served a far more practical purpose in ancient Greece: measuring land. By the seventh century BCE, when they first began to build cities, the Greeks had developed a fascination with the organization of things in space. Surveying and city planning were among the tasks that occupied early Greek geometers. If structuring a city as a way of mapping out social orbs was their goal, why not use the same abstract model to construct a universe? Why not intellectualize about the physical as well as the social universe?

Thales of Miletus, a philosopher from the Greek city in southwestern Asia Minor, was far better known for his astronomical accomplishments than his distant predecessor Homer. He is said to have fallen down a well while stargazing. The beautiful maidservant who helped him out marveled at his knowledge of the heavens despite his inability to notice what lay just beneath his feet. Thales is credited by some with being first to mathematically predict an eclipse, the one that took place on May 28, 585 BCE. In his account of the war in Anatolia (today Turkey) between the Lydians and the Medes, fifth-century BCE Greek historian Herodotus tells us: "As the balance had not inclined in favor of either nation, another engagement took place in the sixth year of the war, in the course of which, just as the battle was growing warm, day was suddenly turned into night. This event had been foretold to the Ionians by Thales of Miletus, who predicted for it the very year in which it actually took place. When the Lydians and Medes observed the change they ceased fighting, and were alike anxious to conclude peace."[9] The story was passed on by the first-century philosopher-mathematician Theon of the city of Smyrna, also in Asia Minor. He quotes fourth-century BCE writer Eudemus of Rhodes, regarded as the world's first science historian for his detailed explanation of the movement of things in the heavens, saying that Thales was first to discover the eclipse of the sun. Science historians still argue over whether Thales was able to explain what causes eclipses, much less actually predict one.[10]

Suppose Thales had actually accomplished the feat of foretelling an eclipse. How might he have gone about it? One way stems from knowledge of the saros cycle, which he could have picked up from Babylonian sources (Lydia was a Babylonian outpost, and there were plenty of astronomy-related texts in the vicinity). I pointed out earlier that Babylonian astronomers used the saros to predict lunar eclipses. Could Thales have connected the 585 BCE event with an earlier solar eclipse? It's a bit of a stretch, but one saros earlier, on May 18, 603 BCE, a total eclipse was visible from South America to Central Africa. It was seen as a partial eclipse in Asia Minor.

On the other hand, why insist on the saros cycle to secure eclipse predictions when there are other eclipse rhythms? It's true that the saros is the shortest cycle that generates recurrent eclipses for a given location on earth. It happens that a rapid sequence of eclipses preceded the one in 585 BCE that Thales could have known about; for example, 84 percent of the sun was covered on the July 29, 588, eclipse for viewers in western Asia Minor. The same was true for the December 14, 587, eclipse (seventeen months later); then, eighteen months later, came, at 90 percent total, the May 28, 585, event.[11] Astronomer Dmitri Panchenko speculates that maybe, in order to calm the frightened Ionians who had just witnessed three near-total eclipses in three years, Thales told them *after* the 585 event not to worry because eclipses happened whenever the moon's orb crossed the sun. It was a natural event and it would happen again, so forget about omens. He might even have issued a timely warning of a future eclipse based on the intervals he'd noticed.

Another Greek eclipse story tells of a major turning point in one of the most significant battles in history. Of the Peloponnesian War, Herodotus wrote, "On the first approach of Spring" the army departed the Lydian city of Sardis and marched toward Abydos; "at the moment of departure the sun suddenly quitted his seat in the heavens, and disappeared, though there were no clouds in sight, but the sky was clear and serene. Day was thus turned into night."[12] Portending this as a sign of defeat, Persian king Xerxes halted his naval fleet as well as his land march, which had been advancing on the Spartans toward their major port city on the Peloponnesian coast, thus signaling victory for the Athenians in the Battle of Salamis. Is it possible that the period of Greek fluorescence that would ultimately lead to the scientific Renaissance might have been forestalled had the Persians won the day? This leads to an intriguing question: did an eclipse of the sun alter the course of Western history?

It all makes for a great story, but one problem with tying this historical account to an eclipse is that no total eclipse actually occurred in 480 BCE, the year of the battle. There was an eclipse visible at Sardis, on October 2, but it doesn't fit Herodotus's description because it was only partial (maximum 50 percent total just past local noon). Two years later, on February 17, 478 BCE, a 94 percent eclipse was visible at Sardis; though considerably darker, it postdates the Battle of Sardis by two years. Persistent archival eclipse chasers note that there was a lunar eclipse on March 13, 479 BCE, but that too doesn't satisfy the description of day turning into night. As far as prediction goes, there is no hint in the historical record that anyone saw any sort of eclipse coming.

Between the lives of Thales and Aristotle (the sixth and the fourth centuries BCE), Athens and Sparta fought the intractable Peloponnesian War. After two decades of battles interspersed with negotiated peace treaties, the Athenian nouveau-riche statesman-warrior Nicias, instigated by a request for help from the colonial city of Segesta in Sicily, took command and laid out an ambitious expansionist course to attack Selinus, a neighbor allied with Sparta. With a force of a hundred warships and seven thousand troops, he set sail for Sicily in 415. But he would need to contend first with Syracuse, the most powerful state in Sicily, also allied with the Spartans. A reinforcing armada joined Nicias, giving the Athenians an initial advantage, but the Syracusans, further bolstered by Sparta, began to turn the tide. Realizing defeat was imminent, the Athenian general planned a secret maritime departure from camp. Thucydides picked up the story: "But after all was ready and when they were about to make their departure, the moon, which happened then to be at the full, was eclipsed. And most of the Athenians, taking the incident to heart, urged the generals to wait. Nicias, who was somewhat too much given to divination and the like, also refused even to discuss further the question of their removal until they should have waited thrice nine days, as the soothsayers prescribed. Such, then was why the Athenians delayed and stayed on."[13]

Nicias's decision to please the gods by holding the army fixed for "thrice nine days," until the next full moon could erase the evil omen, would prove costly. The defeat he suffered as a result of delaying his escape proved to be not only the turning point of the Peloponnesian War but also the decline of Athens as a world power. On eclipse superstitions, Roman historian Pliny wrote: "This alarm caused the Athenian general Nicias, in his ignorance of the 'cause,' to be afraid to lead his fleet out of the harbour, so destroying the Athenians' resources: all hail to your genius, ye that interpret the heavens and grasp the facts of nature, discovers a theory whereby you have vanquished gods and men."[14]

It was Nicias's (and Athens's) misfortune to confront the crisis on the eve of the educated layperson's understanding of the physical nature of the heavens. Only a decade before the Athenian campaign, philosophers like Anaxagoras were teaching that the moon shines by reflected light and eclipses are caused by the contact between the moon and the sun (or the earth's shadow). Meanwhile, Meton, the architect of Athens, had discovered the nineteen-year cycle aligning the phases of the moon with the seasons of the year that bears his name. (Incidentally, he was against the war.) The Athenian public was pivoted on the cusp between superstition and rational inquiry. Opposed to the new rationalism were those who believed strongly in the divine nature of celestial bodies, and many of them happened to be supporters of the war. So what was going on in Nicias's head? Had he heard of scientific explanation? If so, did he reject science because it conflicted with his religion? (He was known to have been a superstitious man who often consulted his personal soothsayer.) Or, as a staunch conservative, did he view avant-garde ideas in general as threats to Athenian society?[15]

These are just a few accounts of eclipses tied to historical events in the classical world. There are others. Thucydides connects a solar eclipse with the Athenian expedition against the island of Cythera, and Roman historians link battles in the Carthaginian Wars and Caesar's crossing of the Rubicon with celestial portents emanating from daytime darkness. But once again the astronomical facts do not always square with the historical account. It all makes me wonder whether those who pursue eclipses via the archive might tend to take historical accounts about natural phenomena a bit too literally. Records of ancient eclipses may not fit the precise time and space standards upheld by contemporary historians, who might assume that what their ancient counterpart describes refers to something that happened where he lives or precisely where he locates his story. We need to remind ourselves that the concept of *precisely* observing and recording data about the natural world, as we do in contemporary science, wasn't widely practiced in times when wonders we label as "supernatural" were shaped largely by the human imagination. How can we know for sure that an eclipse is being described?

Let there be no doubt, however, that the Greeks, at least by late classical times, were well aware of the causes of eclipses and consequently quite capable of mathematically grinding out eclipse predictions, as is evident from what Aristotle has to say regarding what happens when the earth casts its shadow on the moon: "How else would eclipses of the moon show segments shaped as we see them? As it is, the shapes which the moon itself each month shows are of every kind straight, gibbous, and concave but in eclipses the outline is always curved: and, since it is the interposition of the earth that makes the eclipse, the form of this line will be caused by the form of the earth's surface, which is therefore spherical."[16]

Concrete evidence of Aristotle's picture of a geometrical clockwork universe that cranked out eclipses appears a few centuries after he wrote those words. In 1900, a diver in the Aegean Sea off the coast of the island of Antikythera brought up a mysterious mechanism from a wreck dated to 87 CE. About a palm width in size, it consisted of an inner assembly of more than thirty toothed bronze gears moved by a crank on its back.

The Antikythera mechanism, as it has come to be known, is the world's oldest analog computer and planetarium capable of determining the positions of the sun, moon, and planets. (It was also capable of predicting planetary positions and charting the Olympiads.)[17] Initial examination revealed that nineteen rotations of one of its larger wheels coincided precisely with 235 rotations on another, thus meshing together nineteen seasonal years with 235 lunar synodic months (called the Metonic cycle). X-ray analysis later revealed a spiral dial positioned at the back of the instrument that calculated the saros cycle. Symbols

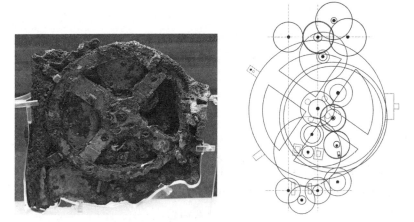

The Antikythera mechanism original remains (*left*) and reconstructed model. The Greeks used mechanical devices to predict eclipses. (Wikimedia Commons)

engraved on the wheel include moon, sun, day, and hour of the night. Other glyphs indicate the type, date, and time of day of the occurrence of eclipses in the eighteen-year saros and fifty-four-year triple saros cycles.

The unique Greek approach to understanding eclipses would not be taken up again for more than a thousand years, as science languished during the Christian era and the Middle Ages that would follow, lending a more religiously centered interpretation to daytime darkness.

7

The Crucifixion Darkness

And when the sixth hour had come, there was darkness over the
whole land until the ninth hour.

—*Mark 15:33*

A number of references thought to refer to eclipse phenomena
appear in the Bible; for example, "For the stars of heaven and their
constellations will not flash forth their light. The sun will be dark,
when it rises, and the moon will not shed its light" (Isaiah 13:10); "But
immediately after the tribulation of those days the sun will be darkened and the moon will not give its light, and the stars will fall from
the sky and the powers of heaven will be shaken" (Matthew 24:29);
"The fourth angel sounded, and a third of the sun and a third of
the moon and a third of the stars were struck, so that a third of them
would be darkened and the day would not shine for a third of it,
and the night the same way" (Revelation 8:12). These descriptions
are fairly generic and difficult to analyze. They all point to signs in the
heavens tied to the celestial bodies that make eclipses happen, signs
portending that something extraordinarily bad is about to occur: the
imminent destruction of the world.

The Crucifixion darkness is one of the most well-known examples of a widely disseminated historical eclipse reference. But is it real

or allegorical? The first three Gospels tell the story. All were written at least forty years after the described event, Mark's version being the earliest. The Gospels of Matthew and Luke likely drew from Mark's account, quoted in this chapter's epigraph:

> Now from the sixth hour there was darkness over all the land until the ninth hour. And about the ninth hour Jesus cried with a loud voice, "*Eli, l'ama sabach-thá ni?*" That is, "My God, my God, why hast thou forsaken me?" (Matthew 27:45–46)

> It was now about the sixth hour and there was darkness over the whole land until the ninth hour, while the sun's light failed, the curtain of the temple was torn in two. Then Jesus, crying with a loud voice, said "Father, into thy hands I commit my spirit!" (Luke 23:44–49)

John (19:31–37) tells a slightly different, lengthier story, but he says nothing about the darkness.

One problem in pinpointing any associated natural phenomenon lies in the uncertainty over the date of the Crucifixion. Historians place it sometime between 29 and 33 CE. Candidates rooted out of the scholarly literature include the November 29, 24, and April 3, 33, events. The first is a bit too early and would scarcely have been visible in Galilee. The second, a lunar eclipse, was considered a possibility because Passover, when the Crucifixion took place, occurs near a full moon; however, the lunar event would not have been visible in the Middle East because it happened during daylight hours.

Close literal readers have pointed out that the biblical "hours" are three or four of our hours long, which seems unusually lengthy for an eclipse; moreover, we know little about how the eclipse was timed and who timed it, and using what devices. Most important of all, the storytellers were not actual witnesses; they gathered the information from earlier second- and possibly third-hand sources. Could the event have been interpreted as the result of a previous prophecy? For example, Old Testament statements in the eighth-century BCE books of Amos and sixth-century BCE Joel predicted that just rule would come to the land in the aftermath of daytime darkness; thus:

> I will make the sun go down at noon, and darken the earth in broad daylight. (Amos 8:9)
> For the day of the Lord is near in the valley of decision. The sun and the moon are darkened, and the stars withdraw their shining. (Joel 3:15)
> I will show wonders in the heavens and on the earth, blood and fire and billows of smoke. The sun will be turned to darkness and the moon to blood before the coming of the great and dreadful day of the Lord. (Joel 2:30)

But a pair of Oxford astronomers who recomputed the date found that the final phase of totality could have been viewed in Jerusalem during sunset (between 6:20 p.m. and 7:11 p.m.), when the eclipse ended.[1] It's possible that a vestige of red coloration due to eclipse could still be visible at the top of the umbral rising lunar disk, in addition to the usual redness caused by dust in the atmosphere. One literal interpreter of natural miracles in the Bible connected the eclipse with the prophetic passage in Joel: "I will show wonders. . ."[2] This strikes me as a bit of a stretch, but it appeals to those who insist on a natural event to corroborate a historical description. Finally, even if there were a matching event, neither of these statements fits with what happens during an eclipse: the sun doesn't go down, and the stars come out and shine rather than disappear.

The birth of the Islamic prophet Muhammad in (570 CE) was said to have been attended by a solar eclipse, that of November 24, 569 CE (approximately 50 percent total in the Middle East and Arabia). The death of the prophet's son Ibrahim also is thought to have been marked by an eclipse. It was supposedly 75 percent total and occurred shortly after the event, on January 27, 632 CE. On that occasion the prophet declared that the sun and moon do not eclipse for any man's death or birth. He asked his followers who viewed the event only to turn to God for prayer.

Another sky event was said to have accompanied the birth of Jesus—the Star of Bethlehem, sighted in the east by the three wise men. What was that star? Conducting research for another project, I came across more than 750 scholarly articles written on the Christmas Star in the twentieth century alone.[3] Suggested natural explanations include a triple conjunction, or close coming together, three times in 7 BCE of Jupiter and Saturn, which might have focused the eyes of the wise men on the west; or perhaps a comet caused them to begin their journey. Comets indeed blazed forth in 5 and 12 BCE. The latter has been traced to an appearance of Halley's Comet, certified in ancient records to have recurred at seventy-six-year intervals as far back as 240 BCE. The range of dates seems to me a bit too early to fit the biblical account. The Christian savior was likely born between 7 BCE and CE 4. Historical events sharing the stage with Jesus's birth, such as Herod's death and Augustus's tax decree, lie along a decade-wide time band. This uncertainty makes it hard to tie the birth of Jesus to a specific celestial event. For example, a Venus-Jupiter conjunction in 3–2 BCE is also compatible with the acceptable dates.

I mention the Star of Bethlehem to raise a bigger question about historical eclipse chasing that troubles me. We have a tendency to assume that a natural event must fit a particular description in the ancient literature. Perhaps the events attending the birth and death of the Christian savior were meant to describe *super*natural phenomena. I know there is little room for miracles in the way most people think today, but the case for an event beyond all scientific analysis still remains a plausible explanation to many religious groups. These people do not necessarily require a scientific explanation in the literature. Had their god been so pleased, he could have created a heavenly event for any purpose, and penetrating the mind of God is no mean task.

"A miracle is simply what happens in so far as it meets people who are capable of receiving it, or are prepared to receive it, as a miracle," wrote theologian Martin Buber.[4] When we try to dismantle an omen in search of its underlying causes, Buber argued, we can lose sight of the meaning it was intended to convey to the true believer who experienced the sign. Some historians have tried to reconstruct natural explanations for the story of Moses's parting of the Red Sea. What combination of wind and water, they ask, could have created an unusually low tide in a shallow bay at just the right time to permit the Israelites to escape Pharaoh's pursuit? But, some theologians counter, the tides in the Gulf of Aqaba are irrelevant to the far more important question of how the children of Israel interpreted whatever happened. For those who followed the Way, that event became an abiding pillar in the edifice of their coming into being as a people.

My immersion in cultural astronomy leads me to think that the search for a literal truth, the precise identification of a celestial event that fits a description in the historical record, is deeply embedded in our contemporary scientific way of thinking. I call it an exercise in scientific demystification, and I have pointed out several examples of this sort of sleuthing in the world of art.⁵ Among them is Edvard Munch's *The Scream*. Completed sometime between 1892 and 1896, this well-known, late impressionist work prominently features a horrified-looking genderless figure clad in black. The hollow eyes, wide-open mouth, and hands clasped over the ears render his scream almost audible. The mood is intensified by confused swirlings of indigo water and a bright yellow-red twilit sky in the background's harbor scene behind the walkway and railing on which the figure stands.

In January 1892 Munch offered a written description of an event that seems to match what he had transferred to canvas: "I walked along the path with two friends. The sun was setting—the sky became blood red and I felt an approaching melancholy—I stopped tired to death—over the blue-black fjord and the city lay blood and tongues of fire—my friends walked on—I stayed behind—trembling with fear—I felt the great screaming [*Geschrei*] in nature."⁶ Experts from the art world have remarked that red and yellow clouds are a climatic oddity in northern Europe, yet they often appear in paintings from that area and time frame. More abstract thinkers say the swirls are visualizations of sound waves or perhaps externalized versions of pent-up energy. One group of scientifically minded sleuths, unsatisfied with such internal, psychological, imprecise explanations, has sought out concrete, external stimuli to explain the background of the painting.⁷ They posed their kind of questions: *Exactly* what did Munch see in the sky? In what *specific* direction was he looking? And *precisely* when did he take that memorable walk?

The scientists interpreted nature's scream to be the eruption of Krakatoa, a volcano in Indonesia, on August 11, 1883. Not only did Krakatoa produce sounds heard a thousand miles away (the cataclysm was recorded on seismographs around the world twice over), but the volcano was also responsible for some of the most dramatic scarlet and crimson sunsets ever witnessed. Three months after the explosion, the *New York Times* reported startled people in the streets gazing at the "bloody red" sky. Many of them believed there must be a great fire in the distance. Extraordinary twilights were reported worldwide up to a year after the blast.

Now, if Munch wrote his description of *The Scream* in 1892, and if he painted it around that time, how can a red sky caused by a volcanic eruption on the other side of the world nine years earlier have been the root cause? Science still can explain the Krakatoa-induced sky, argued proponents of this theory. They found some of Munch's correspondence with friends suggesting that the scream event took place considerably earlier than the artist's famous work. In fact, several sketches survive from as early as 1885. With single-minded persistence, they gleaned clues from the artist's diaries. They ferreted out vital information that led them to the exact spot (a wooded hill in western Oslo overlooking the bay) where, they insist, Munch stood when he was overcome by the volcanically induced sunset.

What intrigues me about such reductive scientific investigations is this culture-based question: Why the deep and persistent interest in pulling away the veil that shelters the naked truth about exactly where Munch was and specifically what he saw and precisely when he saw it? Knowing exactly where and precisely when seem like harmless inquiries; there is no conflict with the deeper issue of trying to sense what an artist feels when he creates a masterwork. But that human issue doesn't seem to capture public attention as much as getting to the bottom line, decoding the mystery.

My little detour via an anecdote about art and reality highlights one of contemporary culture's preoccupations: discovering, to turn Jacquetta Hawkes's phrase about Stonehenge and eclipses, the sort of truth we deserve and desire. What if, instead, we interpret the birth and crucifixion stories as just that, stories, embellishments not to be taken literally, but intended instead to draw the listener's attention to the deep significance of the message being delivered. If the Star of Bethlehem story related by Matthew is *just a story*, that doesn't necessarily imply it is devoid of truth and meaning.

Contemporary theologians explain such references in the gospel as *midrash*, a method for arranging truth through story as old as the Talmud. Quite distinct from reporting a cosmic happening, natural or otherwise, midrash serves to illustrate a religious teaching. The story about whatever celestial phenomenon might have attended the birth and death of Jesus isn't a lie. It reveals what the writers of the Gospel felt to be the truth about a man taken to be their savior. Following the style of the times, they simply may not have been concerned with historical literalism, or at least they were far less concerned than we. Here we need to realize that one of the goals of New Testament Gospel writing is to demonstrate that Old Testament prophecies were fulfilled. Biblical historians are aware that the miraculous accounts referred to were not eye-witnessed and that there were many noteworthy celestial phenomena that occurred in the two or three generations that lapsed between Jesus's lifetime and those who wrote about him that could be used to convey regal attributes to the narrative. The Gifts of the Magi, after all, were a mark of respect for royalty. So, maybe the narrative of Jesus's life is really a story about the good news of salvation, literally the Gospel, and only that. And maybe we shouldn't be too concerned with reading the Gospels in any other way, lest we do violence to their account.

As we immerse ourselves in our quest for specific answers to our own contemporary questions about what goes on in the natural world, I think we need to reflect more on what might have lain in the mind of the painter of the painting or the teller of the tale. Maybe the main point of the Star of Bethlehem story is not related so much to the specific event that set the Magi on their journey. The efficacy of the story lies in getting across the idea that there once was a group of disenfranchised people who lived two thousand years ago. On a night of glorious hope, they were desperately seeking signs of a better future that would surely come with the birth of new light upon the world. Finding *the* star would have had little to do with connecting the believer to the significance of the coming of their savior.

So, was there a star that announced Jesus's birth, or an eclipse that marked his execution? Maybe there was; maybe there wasn't. My own opinion is that there probably was some sort of sky event in either case. It may even have happened after the fact. Verifying the absolute truth about a natural phenomenon portending the Advent or Resurrection is as difficult as finding an optical, electromagnetic explanation for the radiated light in Jesus's transfiguration—or locating the exact spot where Edvard Munch stood when he painted *The Scream*.

Astronomer and historian hold different points of view. They ask different kinds of questions because the knowledge that each holds precious is not the same. While the Advent star and the Crucifixion darkness retain a central place in the stories, once we try to reach out and touch the relevant celestial phenomena, like rainbows' ends, they vanish before our eyes. Like searching for unicorns, the quest for *the* cosmic event may tell us more about what lies in ourselves than in our stars. At the same time, we need not fault astronomers for seeking possible answers to their own questions, such as, could there be physical cause, perhaps an eclipse, embedded in the narrative?

8

Ancient Chinese Secrecy

Here lie the bodies of Ho and Hi
Whose fate though sad is visible,
Being hanged for they could not spy,
Th' eclipse which was invisible.

—*Chinese epitaph*

This Chinese epitaph about Ho and Hi appears in practically every work I have read about ancient eclipses. It refers to the dread fate of two astronomers who failed to forecast a solar eclipse. Evidently they were surprised by the event and did not have time to prepare the customary interpretations of omens and rituals that attend it. (One version of the story has it that the two were drinking on the job.)

The eclipse in question most likely happened in 2137 BCE, during the Neolithic period, when:

On the first day of the last month of Autumn, the Sun and Moon did not meet harmoniously in Fang [the region of our zodiac overlapping parts of Libra and Scorpio]. The blind musicians beat their drums; the inferior officers and common people bustled and ran about. Hi and Ho, however, as if they were mere personators of the dead in their offices, heard nothing and knew nothing;—so stupidly went they astray from their duty in the matter of the heavenly appearances, and rendering themselves liable to the death appointed by the former kings.[1]

The problem with this interpretation is that there is no evidence that the Chinese had the capacity to predict eclipses at that early date. Some experts question whether the statement refers to an eclipse record at all.

Whether or not they could foretell them, Chinese astronomers at this time were extreme eclipse chasers. Their surviving annals record more eclipses than any other ancient civilization. Leaders required advisers to keep a constant watch on the sky for any unusual phenomena, meteors, comets, planetary conjunctions, supernovae, the aurora borealis, and especially eclipses that might portend crises for the dynasty.

Up until the twentieth century, everything in Chinese society revolved around the emperor, who was mirrored in the sky by the fixed Pole Star, the pivot of motion of all things cosmic. The stars of the circumpolar region populated the Purple Palace, the queen, the Crown Prince, the secretary, guards, and other supporters. Four of our seven Little Dipper stars, plus two others, constituted the Kou Chen, or "Angular Arranger," an appropriate term for the celestial determinist. Each functionary in the sky had a terrestrial social counterpart in emperor's palace. For example, the Crown Prince governed the moon, the Great Emperor ruled the sun, the Son of the Imperial Concubine governed the Five Planets, and so on. When the emperor's star lost its brightness, his earthly counterpart would sacrifice his authority, and the Crown Prince would become anxious when his star shone dimly, especially if it lay to the right side of the emperor's. Each follower dutifully circulated about the immovable one. In ancient China, everything that took place in the heavens—in particular, phenomena out of the ordinary—would be analyzed for its omen-bearing content and the result directed exclusively to the royal family. The court astronomer's job was to track every movement of the emperor's heavenly minions and interpret the *meaning* of each event.

Given the demanding duties of the court astronomer, the temptation to cheat, to fabricate eclipse records and interpretations, must have been enormous. Joseph Needham, the greatest sinologist in the West, thought eclipse records emanating from the office of the Grand Scribe Astrologer of the Han dynasty (206 BCE–220 CE) showed signs that court officials manipulated records and fabricated events so cosmic happenings would portend political preferences: "It is probable that a solar eclipse [in 186 BCE] was fabricated in the early years of the Han as a warning to the unpopular Empress Lü (d. 180 BCE) and . . . certain observations of popular solar eclipses were not recorded during the reign of popular Emperor Hsiao-Wen."[2]

Though relatively rare, such fabrications often took place well after the actual event. Under Chinese law it would have been a capital crime to report a "false prodigy" under any circumstances, and despite the lack of reliability of the Ho and Hi anecdote, there *is* evidence of the imprisonment and execution of individuals who fudged the archive.[3]

Ancients the world over watched the heavens diligently, noting the way the stars and planets functioned together like a well-ordered society. *As above, so below*—astrologers believe that what happens in real life is influenced by what takes place in the sky. The drama in the living world overhead creates a parallel plane of existence against which people who reside beneath it can reflect upon and examine their own behavior. Our predecessors personalized nature by recognizing the gods as a form of intelligence with a hierarchy and specialized duties modeled after the structure of human society that underlay the causes of events that transpired in the universe. Inquiring into the wills of animate celestial bodies helped them understand their range of powers, their personalities. Designating stars and planets by name and characterizing them as bearers of messages from beyond lies at the foundation of astrally based religions practiced by most ancient cultures.

The conquest and absorption of one city by another could be foretold by the perceived interaction among the celestial deities to whom the respective cities paid homage. Omens, literally "words from the mouths of heaven," were prescriptions that served as motives and potentials for human action. If things didn't work out, other predictions would be forthcoming, but one must adapt when seeking the signs in nature. As I suggested earlier, of all the signs—the first rains, the budding of trees, the times of arrival and departure of wildlife— those that take place in the sky are the most precisely predictable. This was the turf of the expert astronomer-astrologer. But to judge by the record, in ancient China the astrologer often used other media to assist him in teasing out the meaning of astronomical omens.

Reading lines on foreheads and palms, moles, tea leaves, animal entrails, and tarot cards are among the myriad forms of divination. Underlying the divinatory process is the belief that the key to the future is already present. It is up to the diviner, or "forseer," to apprehend human destiny in whatever medium it happens to be written. One of the earliest for the Chinese was *scapulimancy*, or divination using animal shells and bones, particularly shoulder blades.

The first recorded astronomical inscription in ancient China appears on material made of bone dated to the sixteenth century BCE.

About 100,000 pieces of animal bone and tortoise shell containing an early form of Chinese characters were unearthed from what appears to have been a vast palace archive of the Shang dynasty (sixteenth to eleventh centuries BCE) of Henan province in central China's Yellow River Valley. Later historical descriptions of just how these materials were used, together with their partial decipherment, leave no doubt that divination was the underlying motive for these astronomically related texts, hence their name "oracle bones."

The oracle most familiar to us is the prestigious and authoritative Greek Oracle of Delphi. There a resident priestess, the Pythia, acted as a medium between people and the gods. If you could get through the long queue, you paid her a fee and asked her a question. She would retire to her chamber and go into a trance by inhaling hallucinogenic smoke, recently suggested to have been ethylene.[4] Speaking in tongues, she wailed her response. Her assistant would work it over and cast it in verse form to relay back to the questioner. Some historians view this interpretation of omen-gathering as the imagining of nineteenth-century romantics. If the oracle's prognostications required earthly substances, its efficacy probably had more to do with the interpretive prophets than the mental state of the priestess.[5]

The Chinese divination process was not so different. It would begin by carving questions about weather, crops, and family welfare into polished tortoise belly shells or scapulae (shoulder blades) of oxen. The diviner then placed the glowing tip of a metal rod into one of a series of pre-drilled holes in the shell or scapula, causing the material to suddenly expand. He would then interpret the resulting pattern of radiating cracks. Full documentation of the prognostication was recorded in Chinese characters directly on the piece. For example, on solar eclipses:

1. On day *kuei-yu* it was inquired: "The Sun was eclipsed in the evening; is it good?" On day *kuei-yu* it was inquired: "The Sun was eclipsed in the evening, is it bad?"
2. On day *kuie-yu* it was inquired: "The Sun was eclipsed in the evening; should it be reported to Shang-chia [the earliest of the known Shang ancestors]?"
3. It was inquired: "The Sun was eclipsed . . ."
4. From day *i-mao* to the next day was foggy; three flames ate the Sun and there were big stars.[6]

(Historians have linked this last record to a total eclipse during which the sun's corona, the stars, and the planets were observed.)

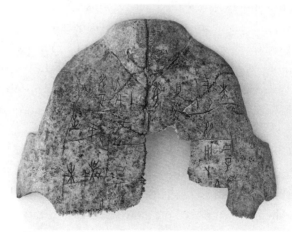

Inscribed tortoise shell from a Chinese Palace archive dated to the sixteenth to the ninth century BCE. Chinese bone and shell engravings that carried eclipse-related texts were used as a form of divination. (Freer Gallery of Art, Smithsonian Institution, Washington, D.C., acquired under the guidance of the Carl Whiting Bishop expedition, F1985.35)

Pending the response, another round of questions might follow: Will the offering of two oxen be sufficient? Will it bring further darkness?

With the vast wealth of astronomical detail recorded on the oracle bone texts, modern historians of astronomy have been able to backtrack regularly recurring celestial events with the computer and match sky phenomena with inscribed dates. We have a fairly precise chronology of events as well as insight into how the ancient Chinese made use of the information they collected. For example, a cache of five thousand pieces of oracle bones excavated in Anyang, Henan province, in 1972 yielded a series of divinations of a single sky event widely referred to as "Ri-you-zhi." Chinese astronomy historian Zhang Peiyu found that all six dates recorded in the inscriptions matched perfectly with a series of solar eclipses visible from the Henan area in the twelfth century BCE, half a millennium earlier than any records of such events that came out of Babylonia and Egypt.

An unusual eclipse event is recorded in a text from around the time of the early Xia dynasty (roughly twentieth to seventeenth centuries BCE): "In ancient times, the three Miao tribes rebelled massively. Heaven ordered them to be killed. The demoniac Sun rose at night. It rained blood in three mornings. A dragon appeared in the temple. Dogs cried in the markets. In summer, there were floods, and earth cracked until water gushed forth. The five grains mutated. The people were thus greatly frightened. Gaoyang thus issued an order in the Dark Palace. Yu himself held the auspicious command from the heaven to attack the Miao."[7]

Astronomer Kevin Pang thinks the sun rising at night fits the description of a total or near total eclipse that took place around sunset. First it gets dark, but once totality ends, the sky suddenly brightens again at dusk, then it gets dark again. This is exactly what I witnessed in Yucatán during the annular solar eclipse that happened around sunset on June 10, 2002. The contrast between the strange *eclipse* twilight and *real* twilight was memorable. Pang found six so-called double-dusk eclipses that matched the period, the best choice being that of April 29, 2072 BCE. Some critics doubt whether the Miao Rebellion statement actually refers to an eclipse, especially because there is no clear way to interpret other statements, such as it having rained blood for three mornings while a dragon appeared in the temple. Nonetheless, I applaud Pang for being less literal-minded than other scientific interpreters.[8]

The first direct evidence of Chinese attempts at eclipse prediction does not appear until late in the eighth-century CE Tang dynasty. One report states that an eclipse foretold did not happen. Modern computations reveal several instances of misses that were actually visible on the same dates in other parts of the world, which may mean that the Chinese, like the Maya, as we'll see in the next chapter, were concerned mainly with eclipse warnings or danger periods.[9] By the eleventh century, astronomers had improved the accuracy of their forecasting and the precision in their reportage, especially for lunar eclipses: "Yuan-chia reign, 13th year, 12th month, 16th day, full Moon. The Moon was eclipsed. The calculated time was the hour of yu [seventeenth to the nineteenth hours]. The eclipse did not (actually) begin until the start of the hour of hai [twenty-first to the twenty-second hours]."[10] During night watches, the time was probably kept on a water clock. We don't know the methods they used, because they were kept secret. By the time the Jesuit astronomers arrived with their Western methods in the sixteenth century, however, they had entered into competitive eclipse prediction with their Chinese counterparts.

At the highest levels, the keepers of Chinese solar eclipse and other astronomical records of observations in China resided in the state observatory. Rather like a mix between the United States' modern-day

National Aeronautics and Space Administration (NASA) and Central Intelligence Agency (CIA), this top-secret institution lay hidden deep within Beijing's Forbidden City. The importance of astronomical observations in the world of Chinese politics made such secrecy, even in the closest quarters, absolutely necessary. On the history of the Tang dynasty (618–907 CE), Joseph Needham documented one directive issued by the king: "If we hear of any intercourse between the astronomical officials, or their subordinates, and officials of other government departments, or miscellaneous common people, it will be regarded as a violation of security regulations which should be strictly adhered to. From now onwards, therefore, the astronomical officials are on no account to mix with civil servants and common people in general. Let the Censorate look to it."[11]

While the progressive Chinese were developing measuring instruments into concrete models of the universe, many of which worked under their own power and consisted of complicated moving parts, Europe was experiencing the Middle Ages. A question long debated by historians of science is: How is it that the Chinese never managed to create out of their new technology an experimental science to test their models, the way the exact sciences had begun to evolve during the European Renaissance? That question is especially intriguing since we know that in many aspects the Chinese possessed technology superior to that of the West. For example, they had already made gunpowder by the third century, developed the clock escapement by the eighth century, and invented the magnetic compass by the tenth.

I believe we can attribute the incredible mass of observational records, as well as the lack of a unifying model for eclipse prediction, to a combination of astrology and strong government secrecy. For centuries, Chinese society was bureaucratically organized. Strong family histories, like those of the Jin dynasty (265–420 CE), contain lengthy chapters on astronomy. These consist mostly of astronomical records, comprising information regarding when and where an object appeared and disappeared, its color, brightness, direction of motion, and so on, and the implications of these data in family affairs. Today, we would scarcely think of mentioning how brightly Arcturus glittered when putting together our family album. I think a related factor is that there was no merchant class in this period; the landed gentry of the early feudal system in China charted the direction of star and state alike. Unlike Plato and Aristotle, both of whom taught in a democratic city-state, Chinese philosophers were intellectuals of the court, and the virtues they instilled bonded agrarian peasant class to ruling warlord and prince. Finally, there was little interstate scholarly communication; put simply, China was not as outward-directed as the West.

Ancient China did not produce scientific models capable of explaining how eclipses occur, and I would argue that cosmic processes, as we now think of them, were not in their purview. For example, the Chinese believed not in a single fashioning of the phenomenal world as we understand it, but rather in a continuous, never-ending creation. Emptiness rather than materiality was their prime focus. Their creator was a maker of *mutations* rather than of things. Past, present, and future for the ancient Chinese had to be viewed in a combined whole that transcended whatever happens to be present in the flesh in the here and now. One cosmology text from 1000 CE illustrates:

> Then there was the Vast Prime [Hunyuan]. In the time of Vast Prime, there still was no heaven and no earth, the empty void had not separated, clear and turbid were not yet divided. In mysterious barrenness and silent desolation, Vast Prime continued for a myriad kalpas [time cycles].

The Vast Prime then divided, and there was Coagulated
Prime [Hongyuan]. Coagulated Prime continued for a myriad kalpas until it reached its perfection. Its perfection lasted
for eighty-one times ten thousand years.

Then there was Grand Antecedence [Taichu]. In the time
of Grand Antecedence, the Venerable Lord descended from
barren emptiness to be the teacher of Grand Antecedence.[12]

I find it difficult to comprehend the source of the power of con-
servatism in such statements about the cyclic nature of time that
would sustain such record-keeping and leave sky concepts so un-
changed through time as Chinese historical records indicate. Keeping
things in the family and confining science to secrecy because of the
value of astronomical knowledge in political prediction certainly in-
hibited the sense of progress we value so highly in Western thought.
When we study astronomies in their cultural context, we can make a
fatal mistake by assuming everybody in the world thinks the way we
do. With notable exceptions in ancient Greece, the notion of educating the public in the ways of progressive science is a habit acquired
relatively recently.

I imagine that, like that of his Babylonian contemporary, the life
of the ancient Chinese astronomer must have been stressful and lone-
ly. He was hardly the free-minded scientific inquirer we portray in the
West, happily working in a modern observatory. The government
nurtured astronomers to accomplish one overriding task: to give the
right time. But what *was* time for the Chinese? As far as we know, it
emerges as an endless accumulation of seasons upon epochs upon
eras. It begins with the mythical creation of life out of the breath of
the divine architect craftsman who first chiseled the world out of the
sky. It continues with the foundation of the original capital city and

dynasty, and traces events down to the present-day festivities celebrating all of these past happenings.

Why bother to look outward to the West for revealed knowledge and new technology when the Grand Antecedence has assured your future in a mandate from heaven? This is the origin of the "China Mystique," which, as historian Gordon Chang wrote, still perplexes us: "Its way of life, language, spiritual beliefs, philosophies, arts, and political order were not just different but monumentally alien. Europeans, of course, encountered many peoples who were very different than themselves, but the Chinese seemed to pose a complex, almost incomprehensible, alternative to the West that defied mastery, either intellectually or geopolitically."[13]

9

Maya Prediction

The face of the sun shall be turned from its course, it shall be
turned face down during the reign of the perishable men, the perishable rulers. Five days the sun is eclipsed, and then there shall be
seen the torch of [k'atun] 13 Ahau.

—*Ancient Maya eclipse prophecy*

More than a thousand years ago, in a dimly lit room of an abandoned suburban residence in Xultun, a ruined city located deep in the
rain forest of northern Guatemala, sits a royal Maya scribe. Legs folded
in the lotus position, pen in hand, he neatly inscribes tiny black markings on a recently stuccoed-over wall, partially covering the once dazzling
red, blue, yellow, and green mural paintings that heralded an important
event in the career of a royal personage who had at one time lived there.

I hadn't realized the importance of the scribe in Maya society until I witnessed Harvard archaeologist Bill Fash's excavation of Pyramid
16 at the Maya ruins of Copan (in Honduras). In a chamber adjacent
to the one holding the jade-encrusted remains of a prominent Copan
ruler lay the bones of a middle-aged male, a set of brushes and an
inkpot at his side. The numbers and hieroglyphs he sketched with
those implements must have mattered a lot to the man he once served,
lying next to him.

While Europe experienced the early medieval period (476–800 CE), the Maya civilization flourished in Central America. In what looks to us like a hostile jungle environment, by the second century CE they had developed highly organized settlements, a complex religious pantheon, and a precise calendar, all of it a consequence of the rise of an elite social class. An abundance of stone stelae, carved with images of the rulers, arranged around long-abandoned plazas of their once majestic cities, Tikal, Copan, and Palenque, gaze back at today's tourists like so many gravestones. Hieroglyphs inscribed in bold relief on the flip side certify their exploits: accession to the throne, betrothal, victory in war, and the taking of captives. Accurate timing seems to have been a top priority for Maya kings and queens. The inscriptions always include complex calendrical dates that tell the precise time of each event, date of the year, position of the planets in the zodiac, correct phase of the moon, and count of days elapsed since the world was created (August 11, 3114 BCE). One text reads: "As Venus replaces the moon, so passed 5 days, 14 months, two years, two scores of years and then it came to pass. . . . He was posted as lord of the accession, Jaguar-Lord of the Palenque lineage. He the Ballplayer, Highest ruler Bird Jaguar, Blood Lord of Palenque. It was then his seating on the jaguar throne in the White House."[1]

Contemporary Maya people, descendants of the builders of America's greatest ancient cities, tell us that the gods created the universe. We don't really know whether their rain, fire, wind, maize, and sun and moon deities were anthropomorphized supernaturals, like Greek Apollo and Athena, or impersonal spiritual forces that made up an animate material world. They may even have been conceived as performers who personify or masquerade as forces of nature. But the gods were forced to do away with their handiwork when the first people they had molded out of clay to help keep the universe in balance were not up to the task. They were unable to speak, so the gods just had them melt away. They tried a second creation, this time fabricating a race of wooden-stick people, but they could not bend their backs to worship their creators. Other failed creations followed, until the race of people made out of corn, the chief Maya staple, proved equal to the task. But the human race is here only on a trial basis. People must make their offerings to the ancestor gods, each class and each individual at the appropriate time, in order to keep things running.

By the eighth century, a handful of brilliant astronomers (contemporary Maya people call them daykeepers) in the service of the ruler developed an obsession for precise timekeeping. They were the seekers of time's harmonic chords, made up of big cycles built out of smaller ones that resonated harmoniously with one another like pleasing tones on a musical scale. If you're a musician, think of Western music's perfect fifth, the interval made up of a pair of pitches with a 3:2 ratio, or the 5:4 measure of a major third.

From what Spanish chroniclers, who first made contact with them following the European discovery of the Americas, tell us, Maya scribes were well educated, especially in mathematics and astronomy. We know of just a few Maya books, so-called codices. Painted on the lime-coated bark of a species of fig tree widely grown in north Yucatán, they were composed just before the Spanish conquest. Believing that the mysterious Maya writing contained only lies of the devil, Spanish priests intent on Christianizing the natives burned piles of them.

The most skilled among the Maya scribes must have been thoroughly acquainted with every page of the codices. Lesser individuals in the school of scribes carried these books as they traveled from town to town, offering advice, giving prognostications about the future,

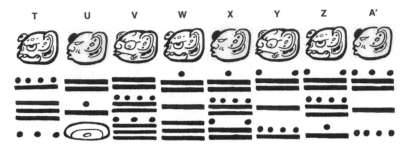

A portion of the restored version of the ninth-century microtext on a wall of Structure 10K-2. (Reconstruction drawing by David Stuart)

and answering questions on the minds of the people: On what day should I commence planting? Should I trade my recently fired ceramics this month or wait until next? Will my child experience a healthy birth? Is there an auspicious day when I might approach my sister to resolve our long-standing dispute? Like their parents and grandparents, the royal scribes who made and cared for the codices were intellectuals, curious about the workings of the natural world and the intricacies of the complex, precise system for reckoning events that linked outer and inner universes their ancient ancestors had created.

Now badly eroded, the finished tiny text in Structure 10K-2 at Xultun is two inches high and seventeen inches long.[2] When archaeologist William Saturno excavated the building in 2011, he and Maya epigrapher David Stuart were able to see the writing more clearly. They recognized a pattern in the right-most three columns, the only decipherable ones in the set. Reading right to left, the large numbers (4,784; 4,606; and 4,429), so painstakingly rendered by the scribe, are separated by intervals of 177 (or 178) days, sets of six months of lunar phases, or semesters. Backtracking to preceding columns by repeatedly deducting 177 (or 178) from each successive entry, Stuart

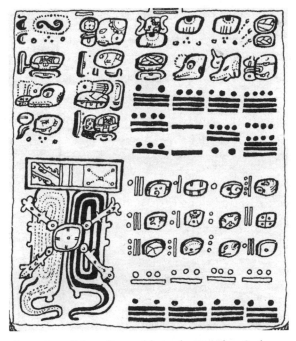

A portion of the eclipse table in the Dresden Codex, a fifteenth-century Maya written document. (Courtesy of Akademische Druck-u Verlagsanstalt Graz, p. 52b)

found perfect consistency with remnants in the surviving number fragments. The text tabulated 162 lunar months arranged in 27 columns of dot-and-bar numbers.

It did not take much to convince us that this unique Maya inscription could be used to predict eclipses. Recall that a perspicacious moon watcher knows another eclipse can't occur in the next month cycle, or the next—you need to wait six (or occasionally five) months before another eclipse can take place. Eclipses can happen only in semesters because the sun and the moon need to be positioned at the nodes, where the two orbits intersect.

Scholars familiar with the scant remnants of the Maya written record were excited by the discovery of the Xultun text. That's because the number 177 is the centerpiece of the eclipse table in the Dresden Codex, one of the surviving documents named after the city in Germany where it resurfaced in the nineteenth century.

You don't even need to read the dot-and-bar numbers in the Dresden Codex to realize that they might have something to do with eclipses. The hieroglyphic text is punctuated by suggestive pictures; for example, there's a serpent-like creature with open jaws devouring the symbol for the sun, along with moon and sun hieroglyphs painted on half-light/half-dark backgrounds.

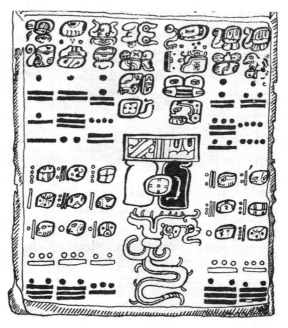

A feathered serpent eating the sun in the Dresden
eclipse table. (Courtesy of Akademische Druck-u
Verlagsanstalt Graz p. 57b)

The red numbers across the bottom of each half-page are the clinchers. They translate to a repeated chain of 177s followed by a 148, the short (five-month) semester.[3] Another series of black dot-and-bar numerals is positioned a few lines above each of these numbers. Part of the sequence depicted on page 52b in the Dresden Codex reads (translating from Maya number notation): 6,408; 6,585; 6,762; 6,939; 7,116; and 7,264.[4] It's easy to see that, if you add the lower number of a given column to the upper number of the previous column, you arrive at the upper number in the next column. For example, 6,408 + 177 = 6,585, and 7,116 + 148 = 7,264. The large black numbers are clearly the totals you would compute by repeatedly adding the intervals below them.[5]

How are the inscriptions penned on a refurbished wall in the ruins of a ninth-century Maya city related to those inscribed in one of their books of prognostications dated six hundred years later? First, the black numerals in both texts are almost exactly the same size. Second, the length of time recorded in the black totals in the Dresden eclipse table is exactly two and a half times the semester sequence inscribed on the wall of 10K-2. Maya mathematicians had a love affair with numbers—the bigger the better—that resonated in the ratio of small whole multiples, like the 405/162, or 5/2, rhythm that harmonizes the two texts. I call it the principle of commensuration.

As I have pointed out, we must resist falling into the ethnocentric trap of assuming others think as we do. Though the 223-lunar month saros appears in the Dresden Codex, it did not play as prominent a role in Maya eclipse forecasting as it did for the Babylonians. The total number of days in the entire Dresden Codex sums to 11,958, or very nearly 405 lunar phase cycles, about 33 years. The cycle that interested them is 11,960 days in length. They chose it because of the overriding ritual requirement of Maya astronomical timekeeping: computational cycles must be commensurate with the most fundamental of all Maya time cycles, the 260-day sacred round. This is made up of the number of fingers and toes on the human body and the number of layers of heaven: 20 times 13. Also, it is close to the gestation period of the human female. So, the 405-month cycle was ideal for the Maya because 11,960 is a whole multiple (46) of 260. Clearly the Maya were dancing to a distinctly different cosmic rhythm than our Western ancestors.

I had long wondered whether the Dresden Codex is a record of eclipses Maya skywatchers had already witnessed, or if it is intended to warn of possible future eclipses. If so, what kind of eclipses? Though scholars are not in agreement over which particular set of eclipses (lunar or solar) was being observed, most are convinced that the information written in the Codex proves that the eclipse table has more to do with warning of possible eclipses. The most convincing clue about the use of the Dresden Codex comes from the introductory page. There we find four Maya dates from around the time the Xultun inscriptions were written down: November 8, 755; November 23, 755; December 13, 755; and October 8, 818 CE. On the first of these dates, a lunar eclipse took place. Fifteen days later (the second date), on the new moon following, a solar eclipse occurred. We think the Dresden Codex is an updated version of an earlier table, or perhaps a series comparable to succeeding editions of one of our farmers' almanacs, updated and intended for consecutive years. My collaborators and I have been examining additional inscriptions adjacent to the lunar semester text.[6] They tell us that Maya astronomers were attempting to project eclipse calculations millennia backward in mythic time to the previous Maya creation era, the one that ended in 3114 BCE.

As work on the Xultun text progressed, archaeologist Franco Rossi began excavating the floor of 10K-2. He uncovered the remains of two high-status individuals wearing the same ornaments depicted in mural paintings on the walls of the chamber. The wall scene portrays three members of a school of scribes. A fourth, more elaborately garbed individual, appears to be conferring with a royal personage on matters pertaining to the celebration of a new year's ritual.

Hieroglyphs accompanying the images identify the former as senior and junior *taaj,* titles bearing the designation of ritual specialists who would possess the knowledge and skills necessary to calculate and write out calendrical tables. Next to one of the individuals, Rossi found a bark beater and a plaster smoother—tools of the trade that could have found use in Maya bookmaking.[7]

XULTUN 10K-2
FIELD DRAWING
H. HURST 2012

Mural paintings on the walls of Structure 10K-2 show junior and senior daykeepers (the black figures on the left) along with their superior, who addresses the ruler (*right*). (Artwork by Heather Hurst, © 2012)

Why were the ancient Maya astronomers so concerned with precision? Knowledge is power, and the ability to decode rhythms in the cosmos that would enable them to anticipate future events would be invaluable to any ruler. The farther back in time you can plant your roots, the greater your legitimacy to rule. With the power of heaven vested in him, the ruler was here to guarantee a perpetual universe, alive with the sacred power of fertility in the earth and in the human body—provided the rituals of sacrifice in debt payment to the Maya gods be made on time. Theirs was a participatory universe.

There's little doubt that the Maya sought to forewarn of possible eclipses because of the disaster they believed threatened them on such occasions. Omen texts that accompany the Dresden Codex read "woe to life," "woe to earth," "woe to the seed"—statements that resonate in our minds with astrology. The pictures in the Dresden Codex look ominous enough, at least to us.

The epigraph for this chapter comes from an early colonial document. It underlines the importance of eclipse watching to the lives of the ruling elite. (The seemingly exaggerated reference to "five days" probably meant the eclipse took place during the month of five unlucky days tacked on to the eighteen months of twenty days at the end of the year.)[8] The link between the polity and ancestor deities is reinforced by inscriptions on carved monuments prominently placed in Maya ceremonial centers.

Contemporary Maya refer to "being bitten, eaten, or blinded" by forces from the underworld to describe what happens during an eclipse. Sometimes the sun god is pictured blindfolded in eclipse texts. Could the agents be the bright planets nearby, especially Venus, which usually appears prominently during totality?[9] Normally thought to be in the underworld when it is invisible as a morning or evening star, Venus suddenly flashes out during an eclipse, seeming to boldly attack the sun in broad daylight. A page of the eclipse table adjacent to a portrayal of the serpent devouring the sun shows a figure diving down from a skyband that supports lunar and solar symbols, his head replaced by a Venus hieroglyph.

Accompanying words, like *xul kin* and *xul haab* ("end of days," "end of years"), refer to the end of time in a broader sense, the end of a long cycle, such as a Maya creation era; each ended with the death of the sun and the instability that usually accompanies any period of transition. The Maya, then and now, believe that the rituals of renewal they conduct aid in the triumph over the forces of darkness.[10]

Foreknowledge of eclipses also had its practical side. The unknown Maya king who reigned at Xultun—like the Copan ruler entombed next to his devoted scribe—would have placed great value in seeking out the most expert skywatcher-mathematicians for his court advisers. Charged with their royal task, the scribes who penned these ancient texts were using the Xultun wall the way my students might use a classroom blackboard or tablet to calculate what would become the finished product, an eclipse warning table designed to fit at the appropriate place in a codex. There it could be consulted to time celestial phenomena that would deliver omens of impending drought, warfare, a marriage alliance—any sign in the sky that might threaten the dynasty. The inscriptions on the wall were vital to guide the future of the ruler. I can imagine another scribe seated alongside the wall text, perhaps the junior scribe, blank book pages in hand, ready to codify the magic eclipse numbers by transferring them to bark paper. No question about it: Xultun's Structure 10K-2 was a made-over workshop—replete with its astronomer-scribes entombed in the floor beneath it.

As I read the handwriting on the Maya wall, I see my own astronomical calculations and assessments in stark contrast to those of my predecessor from another culture. I admire my Maya counterpart, charged with manipulating celestial cycles so that they would rhyme with time loops governed by *ritual* in addition to astronomical concerns. I can only wonder what it would be like to live in and serve a society that merged science with religion.

Let me close with a story that might help convey how far one culture can be from understanding the spiritual and intellectual leanings of another. Diego de Landa, first bishop of Yucatán, tells of an interview with a Maya scribe he conducted during the Inquisition. Belatedly curious about the Maya codices he had hastily burned in a huge bonfire, the prelate naively posed a question about what alphabet the Maya used in their writing system. (Actually, the Maya employed syllabic writing): "Como se escribe 'ah'?" (How do you write "a"?), he asked. Likely puzzled by such a question, the native drew the hieroglyph for "turtle"—*aac* being the closest Maya sound equivalent. Satisfied, Landa continued "Como se escribe 'beh'?"; the perplexed respondent correctly answered the question by drawing a footprint, *beh*—the Maya sound for "road"—and so on. Landa carefully annotated with his own commentary each letter of what he thought were the Maya ABCs. At the bottom of the page of Landa's chronicle appears the entry *ma-in-ka-ti*. Epigraphers have long wondered what that word is doing there. The best explanation I've heard is that, after what must have been an exasperating day, the puzzled interviewee probably became bored with such stupid questions and terminated the interview: *ma-in-ka-ti* means "I do not wish [to]."[11]

10

Aztec Sacrifice

They wrote down and painted the notable things that happened
or were seen in the sky or on the land, such as an eclipse of the sun
or the moon, as well as comets or any other new signs; then when
they began a new fifty-two-year cycle, they put down the house of
the year, and the masters of counting began to record all of the
things worth remembering, wars, the deaths of eminent persons,
earthquakes, famines, numerous deaths, and the like.

—Toribio Motolinía, sixteenth century

Two stereotypes persist about the ancient Aztecs, distant neighbors of the Maya to the northwest who flourished several centuries
after the Mayas' decline: they were sun worshippers and they indulged
in mass human sacrifice. The first is easy to verify. Having set up their
capital city of Tenochtítlan (now Mexico City) on a well-fortified
island in the middle of Lake Texcoco in the Valley of Mexico, the
Aztecs gained independence and commenced a career of imperialist
expansion with a victory over the rival Tepanecs of the neighboring
city of Azcapotzalco in the year they named 1 Flint, or 1428 in our
calendar. This would be New Year's Day in the 365-day calendar expressed in the 260-day sacred round. The number 1 refers to the
first day in the thirteen-number cycle that meshes with the twenty
day names that make up the 260-day (13 × 20) time cycle, the same

cycle used in the Maya calendar. In this instance, Flint is the day name; a loose analogy in our calendar might be a reference to the year 2017 as "Year Sunday 1" because New Year's Day fell on a Sunday.

By 1486, under the leadership of the eighth *tlatoani* (dynasty ruler), Ahuitzotl, the Aztecs had doubled the size of the land under their dominance. Ahuitzotl installed Huitzilopochtli, god of sun and war, as the city's new protective deity. His architects erected the Templo Mayor, or principal temple, so that the sun would rise exactly over its midpoint on the first day of spring, when worshippers would commence their sacrificial rites in anticipation of the rainy season.[1]

So precisely did Ahuitzotl's successor, the emperor Moctezuma II, insist on following his special god of sun and war across the sky that he ordered the upper portion of the Templo Mayor torn down and rebuilt because, as one of the Spanish chroniclers tells us, it was "poco tuerto," or "a little twisted out of line." This must have been a formidable collaborative task shared by architect and astronomer, for only through their successful joint effort could the celestial luminary appease the people by continuing to keep his annual appointment. When we measured and mapped the orientation of the building, I was astonished to see how successful the collaboration turned out to be.[2]

The merging in a single deity of the idea of the sun on the one hand as the source of all energy in the fertile landscape, and on the other as a mandate for war and conquest, proved a clever political move on the emperor's part. He had set as his goal the expansion of a once modest, early fourteenth-century state into an empire that would stretch, less than three generations later, from the far north of the country today we call Mexico to the west coast of Central America.

On the issue of Aztec blood sacrifice, we need to consider the reasons behind such a practice among other cultures of the world.

Generally, one commits a sacrificial act to appease the gods, to divine the future, to strengthen social bonds within the sacrificing community, and to guarantee survival in the afterlife. In the classical world, Plutarch tells of a Spartan mother who raised her sons exclusively for blood sacrifice, that each might die "in a manner worthy of himself, his country, and his ancestors."[3] Told her five sons were killed in battle, another female parent went to the temple to thank the gods. To give a modern example from another culture, political scientist Louis René Beres argues that the jihadist terrorist "ecstatically kills himself and innocent others only to ensure that he will live forever."[4] Suicide is only a short-term inconvenience on the way to life everlasting. The martyrs' actions constitute a sacred act.

What do we know about the relationship between solar worship and human sacrifice in the Aztec worldview, and where do solar eclipses fit in? I think the extent of human sacrifice among the Aztecs has been grossly exaggerated by the Spanish chroniclers, who came over with the conquistadors after Columbus to catholicize the natives as a way of justifying the conquest. The Aztecs' religious tradition portrayed the universe of space-time as the exclusive domain of the gods.[5] The visible world was set aside by them for humans, animals, plants, mountains, stars, and so on, all of which partook of life. It was also the place where supernatural forces mingled with the mundane. The chroniclers tell us that before time began, before the sun was created, native informants told them that the gods needed to offer a sacrifice to make their first creation. The bravest of them leaped into a huge bonfire on the eastern region and descended into the world below, reemerging as the sun. Only then would light be shed on the world; only then could the sun begin its daily course across the heavens; only then could day and night follow one another. But to recover his strength, each day the sun needed to be fed.

Humans were another of the gods' creations. Sustained by the fruits of Mother Earth, they were brought into the world to tend to that task. They were the ones delegated to administer to the resurrected Huitzilopochtli and to offer him sacrifice as their debt payment to their creator-ancestors. But just as the biblical Hebrew god asked Abraham as a test of loyalty to sacrifice his son on Mount Moriah, the currency would be of the most precious kind, their own blood, the blood of sacrifice. So, the people and their gods entered into a contract. Being the recipients of the gifts of the gods: rain, fertility, good health, and, altered by the politics of war, victory in battle, their obligation would be to make offerings and sacrifices to repay the gods in return. This action would forestall the possibility that their deities, being (like people) fickle and capable of anger, might send omens of floods, volcanic eruptions, and military defeat.

Surviving Aztec codices tell us that the sacrifices took place on a regular basis in the framework of the solar calendar, which consisted of eighteen months, each twenty days long, with an extra five-day month at the end of the cycle.[6] These rites usually took place at the completion of each month. More elaborate ceremonies were conducted at the end of a fifty-two-year cycle, the time it takes for a given day in the 365-day year to cycle around and return to its original position in the 260-day sacred round. Aztec historical records seem to suggest that elite citizens regarded it an honor to offer themselves for blood sacrifice. Soldiers and young children, willingly given over by their parents, were considered the most precious offerings. Others included war captives.

The first entry of Cortez into Tenochtítlan in 1519 happened at a time of accelerated Aztec colonial expansion, when political propaganda proffered an increased demand for sacrificial blood. Aztec art of this period is rife with imagery depicting sacrificial flint knives piercing human skulls, scenes showing heart extraction, and blood-soaked stairways ascending temples of sacrifice, excellent grist for the Spanish conqueror to grind out as bad press for the people they were subjugating. As my epigraph implies, extraordinary natural occurrences, especially eclipses of the sun, figured prominently in these sacrificial rituals, as well as in the Aztec version of the interweaving of human and cosmic history.

A relevant multiple-choice question that might help us understand what history meant to the Aztec commoner:

Which event does not belong?
A. The 2011 earthquake in Japan.
B. The 2004 Indonesian tsunami.
C. The 2017 and 2024 eclipses of the sun.
D. The 1980 eruption of Mount St. Helens.
E. The 1968 assassination of Martin Luther King Jr.

The answer is obvious to any American schoolchild, but asked a similar question about current events in the *calmecac,* the premier educational institution in Tenochtítlan, any savvy Aztec youth would be looking for a "none of the above" option. The Aztecs did not differentiate between a history of *things* and a history of *people.* Their worldview didn't separate animate from inanimate.

A newly independent people on the rise to power might well feel the need to revise or reinvent their own such history—and this is precisely what the Aztecs appear to have done. We can't know exactly how earlier records were eventually changed (like those of the Chinese, they were probably fudged for political ends), but we do know that in the final versions that appear in the codices, particular events from the distant past were assigned dates with important positions in the fifty-two-year cycle and that certain types of events, births and deaths of rulers, wars, pestilences, were recorded as occurring in years that had the same name; for example, 1 Flint. Modern historians call these Aztec innovations "like-in-kind" events.[7] At the end of a fifty-two-year cycle, the Aztecs would conduct a ritual of purification. They called it the "New Fire" ceremony. All fires in the hearths of the city's residents were extinguished, and household articles, from cups to dishes to mats, were destroyed. Citizens assembled at the center of the ceremonial precinct, where priests had constructed a great bonfire. At the end of the ceremony, the people bore their portion of the New Fire on torches and returned to their huts; there they reignited the hearth and began time's cycle anew.

The best example I can think of in American history that might help us understand the Aztec concept of history would consist of labeling the sixties as the "decade of assassination," because assassinations of Abraham Lincoln, John F. and Robert Kennedy, and Martin Luther King Jr. all occurred in that decade of the century cycle. Aztec historians, for example, assigned the following events to years named 1 Flint:

1. The beginning of the Aztec migration from their ancient homeland, known as Aztlan.
2. The birth of Huitzilopochtli, god of their city.
3. The arrival of the Aztecs at the site on the island where they were destined to build their future capital city.
4. The enthronement of Acamapichtli, the city's first king.
5. The winning of independence from rival tribes and the beginning of their empire.

Sky phenomena are frequently depicted alongside events of social and religious significance in the surviving Aztec pictorial histories. I can imagine the court astronomer, adviser, and omen bearer to the king, all rolled into one job description, being deeply involved in the process of re-envisioning the official history of the empire. But how did he go about the necessary task of pegging like-in-kind events, such as battles, conquests, famines, and feasts, to natural phenomena, such as eclipses? This is a question that interested me and my colleague, Aztec ethnohistorian Edward Calnek.

We analyzed all the eclipse references in their surviving documents and discovered that Aztec calendar adjustments were frequently geared to the fifty-two-year cycle or one of its multiples, and that visible eclipses were taken into consideration only when they facilitated inserting specific like-in-kind dynastic events into their chronology. These events were based on the accumulation of records, now lost, that were once kept by successive generations of Aztec astronomers, the way Babylonian skywatchers amassed observations recorded in cuneiform that led to their predictions of future eclipses. Like the Maya, in many instances the Aztecs back-calculated events to times in the mythic past, with the goal of more deeply embedding the ancestral roots of the dynasty. An example given by one of the Spanish chroniclers shows how they placed an eclipse event: "In the year 5,097 after the creation of the world, which was 1 Flint, Toltec sages, astrologers, and [masters of] all other arts, assembled in their capital city, Huehuetlapallan, where they examined the happenings, calamities and movements of the sky since the creation of the world, and [accomplished] many other things . . . [and] it had been one hundred and sixty-six years since they had adjusted their years and times with the equinox . . . when the sun and moon were eclipsed and the earth trembled . . . which was in the year of 1 House."[8]

We also discovered that one brilliant Aztec astronomer managed to link the most frequently recorded date for the founding of

Tenochtítlan (Year 2 House, or 1325) with a 99 percent total eclipse that took place one hour before noon on April 13, 1325. As an added bonus, the first day of this calendar year fell two days after the spring equinox, the scheduled time of arrival of the sun god at his station in the middle of the Templo Mayor. Immediately after sundown on the same day, a vertical alignment of four closely gathered major planets, Mars, Jupiter, Saturn, and Mercury, appeared in the western sky, a brilliantly orchestrated natural backdrop for an outdoor religious celebration. Aztec lore tells us that the people made noise during totality to scare away the *tzitzimime* (demons), the stars, and especially the planets that appeared during totality—as they had come to devour the solar deity.[9]

These circumstances, and the selection of a year named 2 House (the year that follows 1 Flint), suggest that Aztec historians chose this particular founding date in their reinvented version of history precisely because the calendar date was reinforced by a spectacular astronomical event. This is the same sort of reasoning behind the choice of cosmic events that accompanied the birth and crucifixion of Jesus. Whether the events actually took place at the precise time assigned them in the story is not as important to them as it is to us.

The most graphic example of like-in-kind astronomical events we uncovered came from the record of a total solar eclipse said to have happened in the year 4 Flint, or 1496.[10] The pictorial representation, which coincides with the near-total eclipse of August 8, 1496, seemed unusually realistic to us. It shows a partially eclipsed sun setting behind a mountain with a close resemblance to the range on Tenochtítlan's western horizon. In other Aztec pictorials, sun disks are usually shown with wedge-shaped bites taken out of them, but this one gives a far more realistic representation of the event. The pictorial is unique because it depicts a clear view of the horizon with the stars shining above it. Modern calculations tell us that eclipse maximum (approximately 96 percent) occurred late in the afternoon. The eclipse would have just concluded as the sun descended behind the mountainous western horizon.

All together we turned up more than sixty-five visible solar eclipses, in various stages, that took place over the period covering the historical record for Tenochtítlan. When we compared our list of eclipses with the ones listed in the Aztec record, it became clear that Aztec sky-watchers didn't necessarily select the most obvious, impressive events to put in their history books. For example, the eclipses they recorded for the years 1504, 1508, and 1510 were not all that spectacular in the

One of many references to eclipses in Aztec pictorial documents, the eclipse of August 8, 1496, near-total in Mexico City, is depicted realistically along the mountainous horizon at sunset, where it was actually visible. (Bibliothèque Nationale de France)

region of the capital city. Being dazzled seems to have mattered less to the Aztec astronomer scribe than noting *when* the eclipses took place: this particular triad of eclipses conveniently bracketed the great "New Fire" ceremony performed in the year 2 Reed, or 1507. Mandated to perpetuate a like-in-kind form of history, the astronomer who kept the calendar instead chose only those natural events that happened to occur in close proximity to important social or religious events.

In sum, the Aztec framework for recording the grammar of time clearly was founded on the course of their solar deity, the chief directive for war and conquest central to the perpetuation of the power elite; court historians and astronomers paid particular attention to pauses in history's narrative, solar eclipses. Selectively recorded visible eclipses emerge as punctuation marks that bracket what we would call *real* historical events, spaced at intervals determined by the fifty-two-year cycle.

Aztec rule was violently terminated by the conquistadors after a bloody two-year siege of the capital. The people who were spared annihilation by sword succumbed to smallpox and other viruses brought over from continental Europe, from which America had previously been hermetically sealed by a vast ocean. But if you take a close look at the Christian churches built on top of and out of the remains of ancient centers of native worship, you'll see vestiges of carved Aztec blocks with images of flowers, water symbols, and sun disks embedded here and there in their walls. We know that the Spaniards made use of the surviving native labor force to erect these centers of a new kind of worship. Maybe it was small consolation to allow the natives to hold on to a few concrete fragments of their once proud cultural identity as they struggled to forge a new one. Or perhaps, as my colleague Eleanor Wake, with whom I had the enriching experience of working on the orientation of Mexican churches, has suggested, it was part of an exercise in the Mexicanization of Christianity?[11]

TOTALITY
Eclipses in the Modern Age

• • •

11

The Rebirth of Eclipse Science
in Islam and Europe

In this year [April 11, 1176] the sun was eclipsed totally and the Earth was in darkness so that it was like a dark night and the stars appeared. That was the forenoon of Friday the 29th of Ramadan . . . , when I was young and in the company of my arithmetic teacher. When I saw it I was very much afraid; I held on to him and my heart was strengthened. My teacher was learned about the stars and told me: "now you will see that all of this will go away," and it went quickly.

—*Abu-Rayhan-Muhammad-ibn-Ahmad-al-Bīrūnī, an Islamic astronomer, remembering a solar eclipse from his youth*

For Arabian astronomers, the science of the stars came second only to religion. They believed it is only in the sky that one can recognize the divine wisdom of Allah. Islam didn't just preserve the science of astronomy in the period between the decline of the classical world and the Renaissance, simply waiting to hand it over to western Europe at the close of the Dark Ages. Rather, Arab astronomers received the gift of geometrical astronomy from the Greeks, cultivated and developed it, and integrated it, along with Persian and Indian astronomical contributions, into their own philosophy. Allah was the begetter of a more personal, human-centered universe than the one they inherited from the classical world. Islam gave back to modern science both algebra and sophisticated pre-telescopic instruments, like the astrolabe, a computer not unlike the Antikythera mechanism for relating time to the position of sky objects. Islamic astronomical tables charted the first visible lunar crescent so that Muslim astronomers could keep track of the appropriate time to worship by the moon's phases rather than via our more familiar solar-based calendar.

An intimate knowledge of the outdoors, both above and below, came quite naturally to the purveyors of this new astronomy. Before they came out of the Arabian Desert in the late seventh century, unified by the prophet Muhammad, the Muslims were a collection of tribes related by family ties. But all had spent centuries navigating by sun, wind, and stars across the vast expanse of desert. A natural interest in numbers, combined with the need to regulate a life of daily worship, accounts in large measure for the careful corrections Arabian astronomers made to previous observations. A Moorish king of Castile, Alfonso X (called "the Wise"), who gave up his crown to turn to sky and instrument, had constructed his own set of tables three centuries earlier. He once said that had he been present at the Creation, he would have made some useful suggestions to Allah about how to order his work! Alfonso's careful tabulations led to the determination of the length of the year, the size of the earth, the change of position of the solar apogee (the point on the orbit most distant from earth), the shape of the moon's orbit, the inclination of the ecliptic to the equator, and the period of precession of the equinoxes. Calculations of all of these earmarks of our spatial universe were thus vastly improved. Without such complexities and details, Renaissance Europe never could have developed the sun-centered theory of the solar system.

Abu-Rayhan-Muhammad-ibn-Ahmad-al-Bīrūnī (973–1048 CE) and Abu al-Hasan'Ali ibn 'Abd al-Rahman ibn Ahmad ibn Yunus

(950–1009 CE) were among a host of Arabic astronomers who observed and recorded eclipses and advanced the Greek astronomical legacy of explaining their causes geometrically. To aid in predicting future eclipses, al-Bīrūnī (quoted in the epigraph) measured the exact coordinates of stars visible during the lunar eclipse of September 17, 1019. Having damaged his eyes while viewing a solar eclipse in his youth, he was among the earliest to warn of the dangers of viewing the sun directly: better you should see its image reflected in water, he said. Unknown in the west until the nineteenth century, ibn Yunus observed more than thirty eclipses of the moon and several solar ones. He compiled a table of geographic coordinates of cities based on his observations, and he wrote "The Great Hakimi" astronomical table, the most accurate handbook published before the Renaissance.

Baghdad and Cairo developed into Islamic centers for astronomical research as early as the ninth century, where professional eclipse watchers meticulously observed and recorded eclipse data. Arab astronomers had two main motives for assembling historical accounts of eclipses: first, to improve their tables for calculating the positions of planets, and second, to benefit future astronomers who might care to go and witness them. For example, on the occasion of the November 11, 923, solar eclipse, an astronomer wrote: "We as a group observed and clearly distinguished it. . . . We observed this eclipse at several sites on the Tarmah [an elevated platform on the outside of the building]. According to calculation from the conjunction tables in the Habash Zij, the middle was at 0;31 [31 minutes] and its clearance at 0;44 hours [44 minutes] calculation being in advance of observation."[1]

Five centuries later, an eclipse chaser traveled all the way to Aleppo from Cairo with the sultan to time the total solar eclipse of June 17, 1433. A marginal note written in his chronicle contains this curious statement, evidently added by someone who accompanied him (the eclipse occurred very late in the afternoon): "The eclipse was dense and it became dark such that we thought that [the time for] the *Maghrib* [Sunset] Prayer had arrived. Then we reckoned that it was still afternoon. I looked at the Sun and found that it was eclipsed and that the eclipse was great. We accompanied the author [the astronomer] to the Great Mosque and prayed after him until it cleared."[2]

Compare these accounts with those drawn from records on the continent, which only spoke of daytime darkening in connection with political turmoil. For example, a twelfth-century English chronicler tells us that in the eclipse of August 2, 1133, the sun was the head of the dragon and the moon was its tail; the event was supposed to have presaged the death of King Henry I. William of Malmesbury wrote: "The elements manifested their sorrow at this great man's last departure from England. For the sun on that day at the 6th hour shrouded his glorious face, as the poet's say, in hideous darkness agitating the hearts of men by an eclipse; and on the 6th day of the week early in the morning there was so great an earthquake that the ground appeared absolutely to sink down; an horrid noise being first heard beneath the surface."[3] Another chronicle backs up the story, except it places Henry's departure from both England and this world in 1135 when: "In this year King Henry went over sea at Lammas [August 1] and the second day as he lay and slept on the ship the day darkened over all lands; and the sun became as it were a three night-old moon, and the stars about it at mid-day. Men were greatly wonder-stricken and affrighted and said that a great thing should come hereafter. So it did, for the same year the king died on the following day after St. Andrew's Mass Day, Dec. 2, in Normandy."[4] But Henry, though he last *departed* England in 1133 (there were no eclipses that year), *died* in 1135.

In her 1900 analysis of the apparent conflict between human and cosmic history, Mabel Loomis Todd concluded that the 1133 event is "not the only labyrinth into which chronology and old eclipses, imagination, and computation, lead the unwary researcher."[5] One explanation for the discrepancy may be that the eclipse and the earthquake that marked his last voyage must have had such an impact on popular belief that they became transferred to his death two years later. (Once the lunar shadow departed the island, it tracked across Europe into the Middle East, passing over Jerusalem, then occupied by Crusaders, in mid-afternoon. It produced four and a half minutes of total darkness, and it was the most recent total eclipse experienced in that city to date.) And in 1349 a story associated with the June 30 lunar eclipse told by Archbishop Bradwardine tells of a witch who had imposed on local villagers. "Make me good amends for old wrongs," she told the cleric, "or I will bid the sun also to withdraw light from you." But the archbishop, who had the good fortune to have studied with Arabian astronomers, responded: "Tell me at what time you will do this and we will believe you."[6]

Beginning about the sixteenth century, eclipse observations of a scientific nature begin to enter historical records on the European continent. "The whole sun was not eclipsed but that there was a bright circle all around," wrote the Jesuit astronomer-mathematician Christopher Clavius, on seeing the annular eclipse of April 9, 1567, in Rome.[7] Johannes Kepler mentions viewing "Red Flames" (the sun's chromosphere, or more likely the corona) around the eclipsed solar disk on October 12, 1605.[8] And the May 30, 1612, eclipse was the first to be viewed through the "optic reed," a telescope tube.

On learning of one report from an observer of the May 12, 1706, eclipse who saw a blood-red streak of light from the sun's left limb for six or seven seconds during totality, John Flamsteed, first Astronomer Royal and founder of the Royal Observatory at Greenwich, concluded in his report to the Royal Society (as it turns out, erroneously): "It infers that the Moon has an atmosphere; and its short continuance, if only six or seven seconds' time, tells us that its height was not more than five or six hundredths part of her diameter."[9] Oxford professor and second Astronomer Royal Edmond Halley, of comet fame, gave this clear description of the corona, which he witnessed as totality approached in England on May 3, 1715: "a luminous ring . . . of a pale whiteness, or rather pearl colour, a little tinged with the colours of the Iris, and concentric with the moon."[10] The prominences were noticed by several observers sixteen years later on May 2, 1733, when observers in Sweden remarked on three or four spots of reddish color near the limb of the moon. Jupiter, Capella, and the stars in the Big Dipper are cited as well.

Notice the way this newfound eclipse reportage focuses more and more on precisely what is going on with the sun rather than in ourselves. Still there remains, even in the eye of the scientific observer reborn, a place for one's feelings. Take William Stukeley's recounting of the total eclipse of May 22, 1724: he began to "feel it, as it were, drop upon us . . . like a great dark mantle." And when it reached totality, the spectacle "was beyond all that [I] had ever seen or could picture to [my] imagination the most solemn." The faces looking skyward around him "had a ghastly startling appearance."[11] As we'll see in the next chapter, the same effect of awe and fear would be expressed by eclipse watchers across the sea in the newfound colonies.

Coming just after the English Civil War, the "Black Monday" eclipse occurred during a time of great turmoil in England. The Council of State that followed the conclusion of the war announced that eclipses were natural phenomena. Visible in London on March 29, 1652, the eclipse fueled public discourse on the legitimacy of astrology. In great anticipation, popular expositions, broadsides, and almanacs were filled with providential prophesying about the coming of the great millennium as a harbinger of change, while many astrologers interpreted the event to be a sign in the heavens of yet another plague like the one that had ravaged the country in 1593. Others portended a great tidal wave would drown the continent and eternal darkness would reign following the eclipse, when the sun would be permanently extinguished.

Political prognosticators warned of "great calamities and mischiefs," the ruin of all monarchy through Europe, especially in England.[12] Knowing the cause of eclipses, other advisers to the court argued that, being natural, the eclipse could have no such ill effects. At the middle ground stood yet another cadre of astrologers who, like those in the modern world, believed that even if the causes of eclipses were not supernatural, this did not necessarily imply that their occurrence would have no influence on humanity.

Public discourse continued, and when the sun rose uneventfully on Tuesday morning, the astrologers defended their flagging reputation. Some blamed it on the Italians: "None of our English Astrologers writ any such thing at all."[13] The eclipse ended up being a public relations disaster for those committed to the "as above, so below" philosophy.

I think Halley deserves as much credit for advancing eclipse predicting as he already receives for foretelling the return of the comet named after him. He was among the first professionals to widely disseminate eclipse information to a popular audience. Halley aimed to elicit public participation in collecting eclipse data. His first opportunity came with the April 22, 1715, total eclipse across central England. Halley published small books and broadsides, or mini-posters, with explanatory one-paragraph snippets that answered questions interested people still ask today: when is it, and where and how can I see it? Halley also put out a map of the eclipse path. In a special "Request to the Curious," he invited interested volunteers to time the beginning and ending of totality, requiring "no other instrument than a *Pendulum Clock* with which most persons are furnished, and as being determinable with the utmost Exactness, by reason of the momentaneous [*sic*] Occultation and Emersion of the luminous Edge of the Sun, whose least part makes Day."[14] Participants were requested to send him the data along with their location. This, he promised, would aid in his calculation of predictions about future eclipses.

But Halley had a second purpose in mind. The accession of King George I to the throne a year before the eclipse had resulted in deep political divisions. Therefore, he wrote, "I thought it not improper to give the Publick an Account thereof, that the suddain darkness wherein the Starrs will be visible about the Sun, many give no surprize to the People, who would, if unadvertized, be apt to look upon it as Ominous and to Interpret it as portending evil to our Sovereign Lord King George and his Government, which God preserve. Hereby they will see that there is nothing in it more than Natural, and no more than the necessary result of the Motions of the Sun and Moon."[15]

Halley was also way ahead of fellow astronomers in exactitude of prediction as well as perspicacity of description of eclipse phenomena. His calculation for first contact came only twenty seconds after he observed the start of the eclipse through his six-foot-long telescope. He expresses the change of color in the sky "from perfect serene azure blew to a more dusky livid Colour having an eye of Purple intermixt."[16] He writes too of "points on the moon's limb," undoubtedly Baily's beads, according to astronomer Jay Pasachoff, for later, Halley concludes that they can have "come from no other Cause but the Inequalities of the Moon's Surface, there being some elevated parts thereof near the Moon's Southern Pole, by whose interpretation part of that exceedingly fine Filament of light was intercepted."[17] Finally, Halley gives one of the clearest descriptions of the corona: " 'Whatever it was, this Ring appeared much brighter and whiter near the Body of the Moon . . . and in all Respects resembled the Appearance of an enlightened Atmosphere viewed from afar,' though he later says he isn't sure about what causes it. He describes the chromosphere as 'perpetual Flashes . . . [which] . . . dart out from behind the Moon in a long and very narrow Streak of a dusky but strong Red Light' on the moon's edge 'that instantly vanished when the edge of the sun reappeared.' "[18]

Following the 1715 eclipse, Halley plotted the actual eclipse path and compared it with the one he graphed based on his calculations. Then he used the result to reconfigure the track of the next eclipse, due in England on May 11, 1724. He predicted the shadow would pass just north of London, about twenty-five miles south of earlier predictions. To solicit public participation in acquiring the timings, he sold copies of the map for one shilling, then reduced it to ten pence (about a 20 percent discount) two weeks before the eclipse.

There were surprisingly few eclipses during the remainder of the eighteenth century in Europe, but by the middle of the nineteenth, the appetite for questing after them was reinvigorated, thanks to a host of spectacular events. Because there were few professional organizations or jobs in nineteenth-century science, the gentleman amateur dominated Victorian eclipse chasing.[19] All the scientific work, including collecting data, was done in the spare time of aficionados, though—thanks to their influence in society—with ample government support. The eclipse of 1870 marked the turning point in planned expeditions in Britain, but, as we'll discover, by the time of the Rocky Mountain eclipse of 1878, the United States would not lag far behind.[20]

As improved rail and ship travel allowed people easier access to exotic places, travel literature and travel agencies blossomed into late

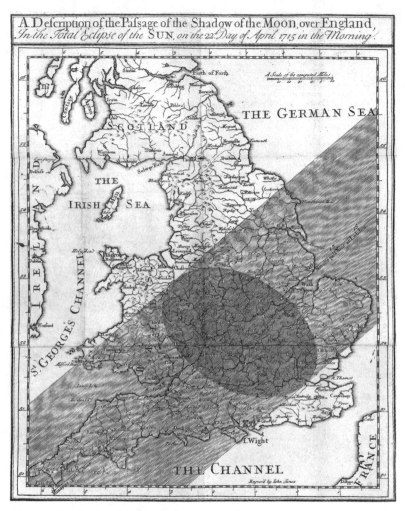

Edmond Halley's eclipse map of 1715, one of the earliest to appear in print. (Collection of Jay and Naomi Pasachoff)

nineteenth-century growth industries. One pamphlet boasted that for every person who journeyed for pleasure in the 1830s, there were a thousand in the 1880s. Whether you sought geological wonders, the great works of classical antiquity, an encounter with the quaint customs in the colonies, or a total eclipse of the sun, the possibility of your journey was a mere question of time and money.[21] En route to eclipse sites, expert astronomers offered lectures to fellow passengers, usually in exchange for reduced travel fees.

Brazil 1893.

Solar eclipse station. Paracuru. Brazil

Typical scene on eclipse day at a busy eclipse expedition site in Paracuru, Brazil, 1893. (Sir Benjamin Stone Solar Eclipse Station, Paracuru, Brazil, 1893. Reproduced with the permission of the Library of Birmingham.)

Given the costs, the earliest apostles of the way of the lunar shadow obviously came from the highest echelons of the aristocracy: lawyers, engineers, architects, and idle, wealthy gentlemen. Their tours, like some of those undertaken by today's storm chasers, combined the didactic and recreational conventions of bourgeois tourism with the productive function of scientific fieldwork, which would later appeal especially to the "work hard/play hard" American ethic.[22]

Descriptions of what nature offered British eclipse chasers in India are awe inspiring, if phrased in a language befitting their status in the world: "There, rigid in the heavens, was what struck everybody as a decoration, one that Emperors might fight for; a thousand times more brilliant even than the Star of India," wrote one observer; another said the sun's corona reminded him "of some brilliant decoration or order, made of diamonds and exquisitely designed."[23]

Assisted by porters to cart their telescopes and auxiliary equipment, British astronomers planned exhaustively months in advance, often needing to put up with dangerous conditions, all for the sake of collecting valuable scientific information in their few precious minutes in the shadow. Some arranged to exercise the duties of their office in comfortable surroundings. Led by noted astronomer Sir Norman Lockyer, members of one expedition in the Moorish region of southern Spain set up headquarters in an observing station on the roof of a café across from their hotel. Before the late afternoon eclipse began, participants enjoyed a popular street exhibit of pictures featuring stars and comets accompanied by dragons and monsters, as one participant noted, "all evidently intended to impress the ignorant peasants, and perhaps deceive them about the great event of the day."[24] Shadow bands and Baily's beads were witnessed, the corona photographed, and then "we all returned to the hotel to tea, eagerly talking over together the wonders of the beautiful spectacle we had seen."[25]

Clouds often got in the way as the crescent sun teased anxious sunwatchers into removing the rain covers on their scopes, only to recede at the very moment they had been waiting for. All that was left to do was to pack up and go home: "It seemed like a death bed, the more so as [the leader] began at once to encase some of the instruments in the black coffin-like boxes," remarked a dejected assistant.[26]

Though they scattered at various points along the eight-thousand-mile-long path of the eclipse of July 8, 1842, astronomers generally associate the event of July 28, 1851, with the first full-blown professional eclipse expedition. Without question it was the first at which an astronomical photograph, a daguerreotype of the totally obscured solar disk, was obtained (at Königsberg [Kaliningrad], Russia). A report of the Royal Society written in the next year lists more than a dozen participants who traveled to Sweden with telescopes and chronometers. They recorded naked-eye observations of the "splendid pink" prominences and produced excellent drawings of the corona.[27]

Scarcely known as a household name among famous astronomers (though there is a crater on the moon named after him), Warren de la Rue (1815–1889) was England's leading exponent of the new technology of astronomical photography. De la Rue was a wealthy Victorian chemist (twice elected to the presidency of the Royal Society of Chemistry) who made his fortune in a printing firm founded by his father. He seemed more interested in the printing process than in editing newspapers (De La Rue Press is still a multinational company). Warren followed in his father's footsteps by developing an extensive scientific background in printing, especially the chemical end of it. The company still retains the "bible" where he filed away the secret recipes for different ink colors used to print stamps for governments around the world.

Young Warren built up his father's business tenfold and spent a good deal of his extra energy and money promoting astronomical endeavors. Eclipse photography entered his life as a result of contact with astronomer-inventor James Nasmyth, who had heard about the white lead coating De la Rue had developed for playing cards, which his firm also manufactured. De la Rue divulged his secret chemical formula and engaged Nasmyth to work on a process for making the pigment. This gave him his first close-up look at reflecting telescope mirrors, and he soon began to grind his own. "You inoculated me with the love of star-gazing," he would later tell Nasmyth.[28] It wasn't long before young Warren invented the photoheliograph, a camera and specially adapted telescope for use in eclipse photography. (The prolific De la Rue also invented an early form of the electric light bulb as well as an envelope-making machine capable of producing 2,700 envelopes an hour.)

De la Rue financed his own expedition to Rivabellosa, Spain, to document the total eclipse of July 18, 1860. Sparing no expense, he pieced together a complete photographic darkroom laboratory at his carefully chosen station in the field. It included a water cistern, a series of sinks, shelves for a dozen chemicals, tables, and a drying apparatus. Adjacent to the darkroom he built a house with a retractable roof for the instrument to give him access to the sky. Over the whole of it he constructed a canvas tent, which he kept wet with a continuous flow of water to lower the temperature in the darkroom during the crucial moments of totality in the heat of the day. To put it all together, he lugged 113 pounds of carpenter's tools, along with lanterns, lamps, several smaller telescopic instruments, a barometer, a thermometer, three synchronized chronometers, glass cutters (to size the photographic plates to fit in the camera), 139 pounds of water—and an undisclosed quantity of wine.[29]

Following a lengthy scientific description of his data collecting leading up to second contact, De la Rue tells how his emotional side took over when he made a sideward glance away from his equipment:

Only a few brief seconds, unfortunately, could be spared from the telescope after totality had actually commenced; but when I had once turned my eyes on the moon encircled by the glorious corona, then on the novel and grand spectacle presented by the surrounding landscape, and had taken a hurried look at the wonderful appearance of the heavens, so unlike anything I had ever before witnessed, I was so completely enthralled, that I had to exercise the utmost self-control to tear myself away from a scene at once so impressive and magnificent, and it was with a feeling of regret that I turned aside to resume my self-imposed duties. . . . And I vowed, that if a future opportunity ever presented itself for my observing a total solar eclipse, I would give up all idea of making astronomical observations, and devote myself to that full enjoyment of the spectacle which can only be obtained by the mere gazer.[30]

Altogether, De la Rue managed to secure forty glass negatives, taken during the three minutes of totality.

He was first to use what is now called the wet collodion process in astronomical photography to increase light sensitivity and detail in the final product. The process was quite trying: he added cadmium iodide to a solution of cellulose nitrate (collodion), then coated a glass plate with the mixture, sensitized it with silver nitrate in total darkness, and exposed it while it was still wet. Following exposure, De la Rue developed it in pyrogallic acid and fixed it in a sodium hyposulfite solution before the collodion dried. Finally, he washed the finished plate, dried, and varnished it. The entire operation, done in a portable darkroom, required several hours of work with caustic chemicals.

Living as we do in a digital age, we can scarcely imagine De la Rue's difficulty in securing a single enlarged quality photoreproduction of an eclipse that happened a century and a half ago. He describes his labors with negative No. 26, the second he obtained during totality, which "gave indications of decay, and on attempting to save it by re-varnishing it, the collodion expanded, and crinkled up so much that,

First wet plate photographic image of totality, taken on July 18, 1860, in Spain by Warren de la Rue. (From *On the total solar eclipse of July 18th, 1860, observed at Rivabellosa, near Miranda de Ebro in Spain* [London: Taylor and Francis, 1862])

except as a record of what was done, the original negative is spoiled. I have protected it from further injury by cementing it with Canada balsam to a second glass plate, but one of the albumen negatives must now supplement it, if more copies are ever taken by direct superposition. . . . As something is always lost in copying, the damage to the original negative is unfortunate."[31]

The observations De la Rue made with naked eyes were as meticulous as those procured with the most powerful mid-nineteenth-century telescopes. We use adjectives like *blood, ruddy, scarlet,* and *pinkish* to describe the color of the chromosphere. To quantify color, he painted a color scale consisting of a dozen or more patches of different hue, ranging from a deep dark red to light yellow, on a paper strip. Once he sighted a prominence, De la Rue compared it with the scale of tints and marked the number of the color that best matched it, in the same three minutes during which he conducted his photography.

Benjamin Apthorp Gould (1824–1896), a proper Bostonian and Harvard graduate, is one among a number of professional astronomers responsible for bringing advanced European technology to America. He studied at Göttingen and worked in observatories both on the continent and in England. Aware of the primitive state of American astronomy, he would become a pioneer in professionalizing and internationalizing his discipline. Gould especially championed the photoheliograph, which he brought over and installed on his telescopes. He took them to Burlington, Iowa, to view the total eclipse of August 7, 1869, on one of the earliest collaborative eclipse expeditions with European astronomers.

12

The New England Eclipse of 1806

Never have I beheld any spectacle which so plainly manifested
the majesty of the Creator, or so forcibly taught the lesson of humil-
ity to man as the total eclipse of the sun.

—*James Fenimore Cooper, 1869*

Between the beginning of colonization and the year 1900, the
United States mainland experienced fourteen total solar eclipses. Six
occurred in the seventeenth century. Like the twin 2017 and 2024
American eclipses, four of them came in pairs. Two paths of totality
intersected in northern Mississippi, then inhabited by Natchez Indians, on July 21, 1618, and October 23, 1623; there are no existing eye-
witness reports. The same occurred for the two western-based eclipses
of November 24, 1677, and April 10, 1679. In the first, the moon's
shadow traversed the Hispanic-occupied southwest; the second passed
from northern California to North Dakota. Sandwiched between the
pair of double headers were the eclipses of November 14, 1659, and
August 22, 1672, both mainland near-misses.

Nature doled out a scant four total eclipses to the mainland in the
eighteenth century—again avoiding the populated colonies. The first
two, on May 22, 1724, and June 3, 1742, were observable mostly to
Native American people of the northwest and the southwest, respec-

tively. It wasn't until June 24, 1778, when the moon's shadow passed across the Gulf coast, Georgia, the Carolinas, and Virginia, exiting the United States at Virginia Beach, that the first colonial American observations of a solar eclipse were attempted, by astronomer and clockmaker David Rittenhouse (1732–1796) of Philadelphia, just days after the British left that city.[1] (Rittenhouse was more well known for documenting another kind of eclipse, the transit, or passage, of Venus across the disk of the sun, a phenomenon that occurs in pairs about once a century.) He was so excited that after he recorded first contact he passed out. But he recovered in time to clock the end of the event, and he used his timings to calculate the distance between the earth and the sun to within 1 percent of its contemporary value. A future president of the United States also witnessed the 1778 eclipse.

Most Americans are unaware that two of their first six presidents were avid natural historians, with a keen eye for astronomy. Historians generally label America's third president as less a scientist and more an intellectual who championed scientific exploration, but if you read what Thomas Jefferson (1743–1826) wrote from the perspective of contemporary natural historians, you will find plenty of original scientific observation in what he has to say.[2] Take his three-hundred-page *Notes on the State of Virginia,* written while on retreat from his governorship of that state during the Revolution. *Notes* is organized as a series of queries, ranging from "Rivers," "Seaports," "Mountains," "Minerals," Vegetables or Animals," to "Laws" and "Weights, Measures, and Money." I found his account of nature's species and phenomena as detailed and scientifically prescient as what appears in the writings of better-traveled adventure-naturalists of the time, like William Bartram. Here, for example, is a portion of Jefferson's lengthy description of one of Virginia's geological wonders, Natural Bridge:

The most sublime of nature's works . . . is on the ascent of a hill, which seems to have been cloven through its length by some great convulsion. The fissure, just at the bridge is, by some admeasurements, 270 feet deep; by others only 205. It is about 45 feet wide at the bottom. . . . A part of [its] thickness is constituted by a coat of earth, which gives growth to many large trees. The residue, with the hill on both sides, is one solid rock of lime-stone. The arch approaches the semi-elliptical form; but the larger axis of the ellipsis, which would be the cord of the arch, is many times longer than the transverse.[3]

Well aware of the problem of using astronomically timed events to calculate longitude, Jefferson wrote, in 1816, that "interrogating the sun, moon, and stars" is the only way we can know the "relative position of two places on the earth."[4] Jefferson was determined to map the exact geographical positions of the courses of rivers that defined the boundaries of the new lands acquired in the Louisiana Purchase. Though he got clouded out of viewing most of the June 24, 1778, eclipse, which was partial on his Virginia plantation, he did write to Rittenhouse asking where he might acquire a more accurate clock to time such future events.[5]

The president would get a second chance, this time successful, on September 17, 1811, when an annular eclipse happened in Monticello. Posting a detailed listing of contact times, Jefferson wrote, "I know of no observations made in the State but my own," and again he complained to Rittenhouse about the lack of a sufficiently accurate timepiece for both astronomical and geographical use.[6]

John Quincy Adams, president from 1825 to 1829, is known largely for having lost a second term in office to Andrew Jackson, who better reflected American popular preferences for things practical and useful. More cerebral, Adams set his sights on intellectual independence from the European continent. An ardent supporter of scientific research in the early republic, his persuasive oratory led to the establishment of the Smithsonian Institution. *Why must we get news of important astronomical discoveries secondhand from the European observatories?* he wondered. Using the metaphor of seeing the light, the eloquent Adams inquired of Congress: "Are we not cutting ourselves off from the means of returning light for light, while we have neither observatory nor observer upon our half of the globe, and the earth revolves in perpetual darkness to our unsearching eyes?"[7]

Enlightened by the techno-scientific instrument revolution overtaking Europe, colonial astronomers responded to America's second encounter with totality, on October 27, 1780, by organizing a rendezvous with the shadow. Harvard philosopher-clergyman-astronomer Samuel Williams laid out a plan to intercept the lunar shadow, due to touch down north of Hudson Bay and pass over Maine at its exit point from the continent in Penobscot Bay. There were two problems: America was in the midst of the Revolutionary War, and the Maine site lay in enemy territory. Undaunted, the American Academy of Arts and Sciences and Harvard University asked the Massachusetts Commonwealth to prepare a ship to transport Williams, three associates, and six students to Penobscot and to apply to the officer in charge of the British garrison stationed there for permission to enter enemy territory.

Science overrode military interests and the British responded positively, but only under two conditions: first, the Americans must depart the territory no later than the day following the eclipse, and second, they were to have no communication with any people living there during their brief stay. Bearing clocks, telescopes, and other scientific apparatus, after an eight-day voyage the ship anchored at the end of a largely abandoned island. Allowing several days to calibrate their clock and set up and test their instruments, America's first eclipse chasers ran into a major setback: though sky conditions were favorable, there was no total eclipse. A report filed later by Williams reads, "The greatest obscuration was at twelve hours, thirty degrees twelve minutes, at which time the sun's limb reduced to so fine a thread, and so much broken, as to be incapable of mensuration."[8]

What went wrong? A Harvard expedition undertaken in 1980 on the bicentenary of the eclipse revealed that either Williams had made an error in his calculations, or he took his geographic coordinates from an inaccurate map of Penobscot Bay; he may also have used erroneous printed astronomical tables. No one can really be sure what combination of circumstances was responsible. On the other hand, one of the benefits of the misplaced choice of location of the expedition was a dramatic sighting of Baily's beads, fifty-six years before Baily himself reported their existence. Astronomer Duncan Steel points out that there are observational advantages to being, as the Harvard astronomers were, nearly fifty miles off the center line of the path of totality, which put them just a few miles from the edge of the moon's shadow. While such a position minimizes the duration of totality, it makes the events just before second and just after third contact far more dramatic. Because they depend on the light just reaching your eyes around the crinkly edge of the moon, both the diamond ring and Baily's beads can last up to ten times longer if observed near the fringe of the eclipse path, where they appear to run quickly around one edge of the moon. (An added bonus is that the elusive shadow bands are easier to see and may last up to five times longer. More important still is the opportunity to get a longer glimpse of the flaming red prominences of the chromosphere, the most colorful feature of an eclipse. And again, because the sun and moon move on the same tangent line near the edge of the eclipse path, observers get up to a minute and a half to catch its spectrum.)[9]

A decided minority in the early years of nationhood, scientifically educated Americans understood the rational explanations that lay behind colossal celestial spectacles, like meteor showers, comets, aurorae, and solar eclipses. They believed the movement of things in heaven followed the laws of nature. Astronomy was among the sciences that offered natural phenomena as public events. To take advantage of these circumstances, and imitating the European tradition, articulate scientific practitioners were eager to make citizens aware of the process of scientific discovery, via public forums and lectures, readable booklets, and articles in popular magazines (*Godey's Ladies Book,* begun in 1830, frequently published science-related articles intended for the education of women). Broadsides served as another popular form for transmitting the latest scientific discoveries.

It took a while for public lectures to gain traction on the under-educated, less urbanized western side of the Atlantic, but a new sense of cultural nationalism incorporating President Adams's desire for intellectual independence from Europe eventually helped America build its own scientifically educated community.[10] Especially between the Revolutionary and Civil Wars, the lecture circuit via the lyceum, a set of local organizations that provided public educational programs and talks, developed into a successful mode of evening entertainment for conveying useful knowledge to the citizen. The lyceum movement featured presentations by the top scientists in their fields along with other notables: paleontologist Louis Agassiz of Harvard; James Smithson, principal donor and founder of the institution that bears his name; and John Quincy Adams. Benjamin Franklin is said to have acquired his keen interest in electricity from attending a public lecture on chemistry.[11]

Though the early nineteenth century would produce the first canon of eclipses (in 1816 in Europe), as well as America's first published maps of eclipse paths (for the eclipses of 1831 and 1834), this was still a time when popular culture, especially in America, was heavily invested in the inherent supernatural power of eclipse prophecy.

The British mathematician-theologian William Whiston, Isaac Newton's successor as Lucasian professor at Cambridge, wrote extensively about eclipses as the divine signal for the end of centuries of religious persecution. Following in his footsteps, the nineteenth-century popular American historian Richard Devens described this frightening eclipse scenario: "Almost, if not altogether alone, as the most mysterious and as yet unexplained phenomenon of its kind . . . stands the dark day of May 19, 1780—a most unaccountable darkening of the whole visible heavens and atmosphere in New England."[12] For Devens, what people in rural Massachusetts saw that day bore little resemblance to the description of what the eyes of Rittenhouse and Jefferson had perceived. He continued:

In the morning the sun rose clear, but was soon overcast. The clouds became lower, and from them, black and ominous, as they soon appeared, lightning flashed, thunder rolled, and a little rain fell. Toward nine o'clock, the clouds became thinner, and assumed a brassy or coppery appearance, and earth, rocks, trees, buildings, water, and persons were changed by this strange, unearthly light. A few minutes later, a heavy black cloud spread over the entire sky except a narrow rim at the horizon, and it was as dark as it usually is at nine o'clock on a summer evening. . . .

Fear, anxiety, and awe gradually filled the minds of the people. Women stood at the door, looking out upon the dark landscape; men returned from their labor in the fields; the carpenter left his tools, the blacksmith his forge, the tradesman his counter. Schools were dismissed, and tremblingly the children fled homeward. Travelers put up at the nearest farmhouse. "What is coming?" queried every lip and heart. It seemed as if a hurricane was about to dash across the land, or as if it was the day of the consummation of all things.[13]

Especially in rural America, nature was regarded as disorganized, chaotic, and wild; hence the term *wilderness*. Once they became better acquainted with the vast forest that surrounded them and to a degree, tamed it, people began to change their view—they came to regard natural features and events as a source of religious inspiration. One of the major thinkers of this idealization of nature was the Irish political philosopher Edmund Burke (1729–1797), who wrote a treatise on the origin in nature of ideas of the sublime.[14] Burke characterized a way of looking at natural events that begins in fear, then becomes transformed into a feeling of awe at the power of god manifested through his creations. Reaching this state of sublimity had the effect of deepening one's faith. Burke theorized that the thrill of this initial fear, which has the power to destroy us, is reminiscent of the apocalypse at the end of time, which gives way via the feeling of the sublime to the eternal bliss that is sure to follow. But you can only attain the state of sublimity by experiencing nature in its most threatening form, such as when you witness lightning and thunder, the destructive power of storms and ocean waves, or a total eclipse of the sun.

Historians trace Burke's thinking about the sublime to travel writing that had become popular in Britain earlier in the nineteenth century, books and essays that recounted ventures into the mountain wilderness. These works offered an alluring alternative to what many conceived as a decaying modern society, a brief respite to liminal space. Acceptance of the idea of the sublime in Burke's early Romantic period treatise on aesthetics took firm hold in America in the first decades of the nineteenth century, when a period of rapid growth fueled intuition and imagination in art and literature as citizens in the eastern corridor began to penetrate the wilds on their western periphery. We find it in the dashing stories of James Fenimore Cooper and Washington Irving and in Thomas Coles's alluring Hudson Valley School paintings.[15] But early American Romanticism was not without its detractors. Author Theodore Dwight, brother of Yale president Timothy, termed the tendency to use one's emotions to understand natural events a mere fad among less educated members of his social world. Why would anyone care "to see the world through a fancied medium" when, with knowledge based on discipline and instruction, you can "view it *as it is?*" (emphasis mine).[16]

The romantic interpretation of exotic astronomical phenomena in early eighteenth-century America was reinforced by the Second Great Awakening, a religious revival movement rife with end-of-the-world prophecy that began to take root in rural and frontier communities. Philosopher George Santayana harshly judged this period, especially in New England, as "an Indian summer of the mind," when poets, historians, orators, and preachers broached native subject matter though they "lacked native roots and fresh sap because the American intellect itself lacked them. Their culture was half a pious survival, half an intentional acquirement."[17]

Itinerant prophets interpreted unusual natural events as announcements of physical catastrophe, the doom of fiery judgment cast down upon sinners. Others read signs in the heavens as portents of the gradual ascent of an age of God's mercy and a last chance at redemption. But theologian and prophet alike agreed: something *big*, or at least the beginning of it, lay just over the horizon; biblical prophecies, however interpreted, were on the verge of being fulfilled. For example, pastor William Miller, founder of the forerunner of the Seventh Day Adventist Church, precisely calculated the date of the Second Coming from biblical scripture: October 22, 1844.

An extraordinary number of terrestrial disasters had led up to the Second Great Awakening, thus fueling Miller's movement. The great Lisbon earthquake of 1755, the worst ever recorded, had been widely reported as a sign of the times that the world surely was about to end. Shocks were felt as far north as Greenland and west into the Caribbean area. Thousands were killed in tsunamis measuring up to sixty feet. Additionally, there were climatic disturbances in the northeast, such as the "year without a summer" (1816), when repeated frosts decimated crops, and flash floods in 1811 and 1826. An outbreak of meningitis in 1810 killed six thousand people, and cholera, originally carried by European immigrants, reached epidemic proportions in New England and New York in 1832.

How to account for so much misery cast down from heaven on a people so strongly devoted to their God? "Surviving disaster is not enough. Disaster must also be fitted into an orderly world view to which moral purpose responds," explained historian Michael Barkun.[18] The more attention given to the problem of moral order, the greater the reception to alternative ways of framing it; so, the greater the tendency to question orthodoxy. Barkun read the Great Awakening as a circular process: believing themselves to be punished for their sins, the deeply devout redoubled their efforts to rejuvenate themselves. In an atmosphere of religious enthusiasm, their shared attitude of victimization by disaster led to more religious revivals. It all happened at a time when many marginal members of society, their agricultural productivity decimated by the harsh rule of nature, were beginning to migrate north and westward into the wilderness.

In competition with the science education lecture circuit, doomsday preachers sowed their seeds of prophecy all across the northeastern United States and Canada. They employed camp meetings, a Presbyterian innovation, choosing sites that were easily accessible by rail to urbanize their efforts. One New Hampshire lecture was attended by 15,000. Revivalists issued special newspapers, books, pamphlets, and periodicals. They searched out past records of astronomical prophecy—for example, the darkening of the sky (later attributed to forest fires) on May 19, 1780 (five months before the eclipse), spectacular northern lights displays, especially in 1827 (a peak year in the solar magnetic cycle), and the total eclipse of the sun in 1806.

A total eclipse of the Sun is one of the most engaging and uncommon phenomena, which Astronomy ever presents to our view. A central eclipse of the Sun will happen in some part of the earth in the course of every year; but it is but very seldom that a total eclipse is seen at any particular place.

We had a favorable opportunity for observing one of these eclipses on Monday last. It was nearly total in this place; the weather was fair and the air serene and clear.[19]

So the *New Hampshire Federalist* reported on the five-minute total eclipse of June 16, 1806, the first of four that would visit the mainland in the nineteenth century. The lunar shadow swept from coast to coast, entering the continent at the northern end of Baja California. After passing through the sparsely populated territory acquired three years earlier in the Louisiana Purchase, the wave of darkness sped through Ohio, Pennsylvania, upstate New York, and New England, finally departing off Cape Cod. How to interpret God's intent based on what we witness during a total eclipse of the sun? Two competing religious viewpoints addressed that question in 1806. They would be played out in sermons preached before and after the great New England public event.

Age-old "revealed theology," based on personal experience and biblical scripture, tells us directly of God's presence in the universe. But beginning in Great Britain in 1802, Unitarian William Paley (1743–1805) published one of the most important books of the nineteenth century: *Natural Theology, or Evidences of the Existence and Attributes of the Deity Collected from the Appearances of Nature.* There he laid out his philosophical arguments for God's existence and his role in the universe based on the exquisite design of nature. It was Paley who originated the familiar analogy of God as a kind of watchmaker who fabricated his instrument to perfection, then let it run on its own, a marvel to be appreciated and understood both aesthetically and morally. American works on astronomy began to reflect the newfound bond between God and nature. For example, Thomas Dick, in his widely circulated 1844 book surveying astronomical discoveries, wrote, "We ought to connect our views and investigations with the supreme agency of Him who brought them into existence."[20] And in a more philosophically oriented 1870 work on parish astronomy, Connecticut pastor E. F. Burr suggested that "we may say that the very nature and circumstances of Deity would *demand* of him, [that] he should create a generally steadfast, law-abiding universe."[21] What transpired in two New England churches on the Sunday following the eclipse day in 1806 ideally captures the conflicting interpretation between revealed and natural theology.

Asa M'Farland (1769–1827) was a prolific writer, publisher, and editor-proprietor of the *New Hampshire Statesman*. Among his most influential works in print was a lengthy diatribe against Christian heretics: those of the Jewish and "Mahometan" faiths, and those who believe in natural religion.[22] In a sermon preached in Concord, New Hampshire, M'Farland reminded parishioners of the melancholy gloom attending the Crucifixion darkness. He denied that the eclipse could have been caused by the moon passing in front of the sun, claiming (erroneously) that the moon was full on June 16 and therefore could not have caused the darkness. Rather, M'Farland related the true meaning of the eclipse: "God saw fit to give an awful memorial to a guilty world, by suspending the light of the sun, at once to show mankind the criminality of sin, and the dignity of the Personage who was then suffering to make atonement."[23]

In his eclipse sermon on revelation, M'Farland denies the laws of physics. Matter is not capable of motion on its own. It needs some power to make it move. And how could a universe come into existence by chance, when mortals are capable of calculating and precisely predicting the eclipse in advance: "Can anyone be so stupid?"[24] The momentous and awesome suspension of light ought to remind us of how dependent we are upon the Lord. Imagine what it would be like if, with a breath of his mouth, God were to suspend solar light and heat permanently. The eclipse is also a reminder of how inconsequential we are before God: "as but a drop of the bucket."[25] For M'Farland, the eclipse is just a little taste of nature's ultimate dissolution, which will happen on Judgment Day: "Will the heavens and the earth, which we now see, be involved in one common ruin? Will every earthly foundation be taken away? Where will any find security? . . . Though the light of the natural sun will be extinguished; yet in the New Jerusalem, the City of God, and the abode of saints made perfect, there will be no need of the sun, nor of the moon, to shine in it."[26] But, M'Farland went on, God will light the way. And so the call to action: "While time is hurrying us on toward the eternal world; and nothing but the precarious thread of life separates us from our irreversible condition; we are most importunately urged to secure the friendship of God, against the day of account."[27]

That same Sunday, a much older Reverend Joseph Lathrop (1731–1820) preached a different sort of sermon to an admittedly more urban audience at his Congregational church in West Springfield, Massachusetts. A descendant of New England pioneers, Lathrop served as pastor in that church for sixty-two years. Printed versions of his sermons reached a wide audience well beyond his denomination. "The best works in my library are those of the Reverend Dr. Lathrop," wrote an admiring Episcopalian minister.[28]

Lathrop agreed with M'Farland on two points: he told parishioners that when he saw the world covered in darkness the week previous, it called to mind what happened at Jesus's crucifixion. This caused him to contemplate what will happen on the final judgment day, when night will become permanent for those who have sinned. Then his words veered sharply metaphorical: "The darkness of an eclipse the prophet improves, though not as an omen or an emblem of national judgments. He warns his people that a metaphoric political darkness may overspread their country, in the same surprising manner as literal darkness in a solar eclipse falls on the unsuspecting earth."[29] While stressing the spiritual analogies surrounding the eclipse, Pastor Lathrop firmly embraced the rational scientific explanation of what had just transpired:

We have reason to rejoice in the progress, which has been made in the sciences, and particularly in the noble science of astronomy. By this we are freed from many superstitious terrors, which, in the dark ages of the world tormented mankind. These rational sentiments are far more conducive to piety and virtue, than the terrors of that superstitious ignorance, which views every comet flaming in the sky, every obscuration of the sun at noonday, every failure of the full orbed moon and every unusual noise bursting from the clouds, every strange appearance in the heavens and in the earth, as awfully portentous of some dire, but unknown calamity.[30]

Finally, Lathrop added another symbolic moral component to eclipse watching: the temporary darkness of an eclipse is followed with cheerful light that "shines more and more unto the perfect day. This is a natural emblem of moral change, in which a soul is brought out of the darkness of sin and into the marvelous light of purity, pardon and peace."[31]

Lathrop's sermon on the eclipse of 1806 strikes me as an early attempt to straddle the divide between science and religion. Like the early writers on parish astronomy, Lathrop struggles with how to deal with incorporating the newly acquired scientific way of knowing nature into his faith by explaining that God has taught us a kind of morally based astronomy to redirect our prescience of celestial events. As clouds foretell a shower and winds a storm, an eclipse presages, however distant in time, a judgment by God. And though we may have the capacity to precisely predict the path of an eclipse, with all of our mathematical and astronomical skill, we shall never accurately read the mind of God.

A few hundred miles west of the New England preachers, in a tiny town in remote upstate New York that bears his family's name, a young man also witnessed the great American eclipse of 1806. James Fenimore Cooper embraced a romantic account of it in a delightful literary essay he wrote later in life. Admitting to the tinge of selfishness he felt in being favored above those who happened to live north or south of Cooperstown, Cooper describes citizens awaiting the eclipse routinely going about their busy tasks, women carrying their water pails, the blacksmith clanging away on his iron, carpenters hammering nails, and teamsters rolling their loaded wagons along the dusty main street flanked by kids at play. Powers of reason aside, Cooper recalls that there were a few "philosophers of the skeptical school . . . who did not choose to commit themselves to the belief in a total eclipse of the sun . . . simply because they had never seen one."[32] When one of them left town on a business trip that morning, he told his wife he wasn't running away from the eclipse, but if the scientific prediction held true and it caught up with him, he'd be man enough to admit it had happened.

Cooper's precise description of first contact, at 9:30 on an extraordinarily warm, cloudless morning, resonates with Edmund Burke's experience of the sublime in nature:

I had scarcely returned to the family party, left on the watch, when one of my brothers, more vigilant, or with clearer sight than his companions, exclaimed that he clearly saw a dark line, drawn on the western margin of the sun's disc! All faces were instantly turned upwards, and through the glasses we could indeed now see a dusky, but distinct object, darkening the sun's light. An exclamation of delight, almost triumphant, burst involuntarily from the lips of all. We were not to be disappointed, no cloud was there to veil the grand spectacle; the vision, almost unearthly in its sublime dignity, was about to be revealed to us. In an incredibly short time, the oval formation of the moon was discerned.[33]

That got everyone's attention. Hardworking transplanted Yankees dropped what they were doing, began to amass in the street, and cast their eyes upward.

During the long wait for second contact, Cooper's attention was attracted to a prisoner, shackled and straining to peer upward through the bars that confined him in the local courthouse in a vain attempt to see what was happening above. The man, a schoolteacher, had been sentenced to die for having severely beaten his child niece to death because of his impatience over her inability to pronounce certain words; it was the young town's first capital offense. The criminal had confessed and repented many times over for his deplorable actions. Due to be sent to the gallows, he was granted a short reprieve so that he might view the eclipse. Imagine, writes Cooper: here was a man imprisoned and denied the light of day for more than a year, now enabled to bear witness to a moment of the most exotic kind of sunlight imaginable: "The wretched criminal, a murderer in fact, though not in intention, seemed to gaze upward at the awful spectacle, with an intentness and a distinctness of mental vision far beyond our own, and purchased by an agony scarcely less bitter than death. It seemed as if, for him, the curtain which veils the world beyond the grave, had been lifted. He stood immovable as a statue, with uplifted and manacled arms and clasped hands, the very image of impotent misery and wretchedness."[34]

At 11:10 a.m., the 4^m45^s of totality began as the somber yellowish color of the landscape faded rapidly, so unlike the onset of an ordinary evening. And then the stars came out.

Cooper likens totality to two ships passing in the sea, sailing on opposite courses. We think of the sun and moon as a source of light, but here we sense the positive material nature of the moon—a "vast mass of obvious matter" interposing itself between earth and sun.[35] At 11:12 the writer's eyes capture the rays of light that stream out of the corona. When it is over, he tells us, women in the streets have tears streaming down their cheeks. Finally, as he struggles to express the instinctiveness of his feelings on witnessing an awesome event transpiring in the sky, he reveals his inner sense of sublimity:

> The peaceful rainbow, the heavy clouds of a great storm, the vivid flash of electricity, the falling meteor, the beautiful lights of the aurora borealis, . . . these never fail to fix the attention with something of a peculiar feeling, different in character from that with which we observe any spectacle on the earth . . . an instinctive sense of inquiry, of anxious expectation, akin to awe . . . whispering to our spirits, and endowing the physical sight with some mysterious mental presence . . . a clearer view than [I] had ever yet had of the majesty of the Almighty, accompanied with a humiliating, and, I trust, a profitable sense of my own utter insignificance. That movement of the moon, that sublime voyage of the worlds, often recurs to my imagination, and even at this distant day, as distinctly, as majestically, and nearly as fearfully, as it was then beheld.[36]

Not far away, but centuries before James Fenimore Cooper observed the total eclipse of 1806, Native Americans witnessed a signal event that would achieve great significance in their culture. The League of Haudenosaunee, or Iroquois, has long been regarded as one of the world's oldest democracies, a forerunner, along with the governments of Iceland and Switzerland, of our own system of representative democracy. It began as a non-aggression pact among the Indian tribes of upstate New York—the Mohawk, Oneida, Onondaga, Cayuga, and Seneca—prior to taking the form of a political confederacy.

An old Seneca legend tells of an eclipse that happened as they consented to join the League: "a darkening of the Great Spirit's smiling face" that was witnessed when the corn was receiving its final tillage.[37] Believing an end of June to beginning of July date would fit best seasonally, historian Paul Wallace sought out eclipses just before European contact visible from Gonandaga, now Victor, New York, a small town southeast of Rochester and west of Cooperstown, where the great council meeting was thought to have taken place.[38] Wallace targeted the eclipse of June 28, 1451, which was 96 percent total there, as the best fit.[39] There things stood, until historian Barbara Mann and astronomer Jerry Fields came up with a date three hundred years earlier, August 31, 1142, which, though 96 percent total, doesn't quite fit with the final tilling of the corn. Notwithstanding, Mann thinks Wallace failed to widen his eclipse search window to accommodate earlier dates because of the academic politics of the times over the invention of wampum, which many white scholars insisted was European. The early date receives support from oral accounts of the history of family lineage, which puts the League's initiation roughly at 1090 CE. Unfortunately, one of the most contentious issues in the scholarly community involves the reliability of oral history, which is largely dismissed by those who insist on admissible evidence only on the basis of the written record—that is, European accounts. There is a graphic record to back up the early date as well. The "Stick of Enlistment," a cane-like stick, marks the number of Tadodahos, or speakers of the Iroquois Confederacy; there were 145 of them, with an average length of appointment of a little over six years, not unreasonable for a democracy (there have been a total of forty-five United States presidents over 240 years, each serving an average of 5.3 years).[40]

Very recent archaeological findings further support an early date, even earlier than the twelfth century. Archaeologist Jack Rossen obtained mid-tenth-century radiocarbon dates from his excavations at a Cayuga Indian site that contained longhouses without palisades, which implies it was built in a time of peace. This supports what contemporary Haudenosaunee have been saying: that the Iroquois Confederacy is over a thousand years old. As a result, Rossen proposes the August 18, 909, eclipse, which was 90 percent total in Gonandaga, as the event referred to in the historical record.[41] That raises the average length in office of a speaker to about that of a two-term United States president, and it makes the Iroquois Confederacy the oldest continuously functioning democracy in the world.

Euro-Americans were not alone in early nineteenth-century eclipse prophesying. In the aftermath of the Revolution, the British had ceded land west of the Appalachians to the new Union and settlers began pouring over the mountains and, via the Erie Canal, into Ohio and Indiana territories. Under stress to adapt to sudden social and economic changes, the Shawnee were among the most vehement opponents of the incursion. With the leadership of their chief Tecumseh (1768–1813), they played a role in uniting local tribes against the white invaders. Tenskwatawa (1771–1836), the chief's brother, who had been strongly influenced by the Shaker Church, was well acquainted with Doomsday warnings from signs in the heavens. He had declared himself an adept among the allied tribes. Tenskwatawa (his name translates as "open door") was famous for having performed the miracle of his own resurrection. He had unwittingly overdosed on intoxicating herbs and remained in a state close to death, when he suddenly awoke two days later, just as he was about to be buried.

During his absence from consciousness, Tenskwatawa claimed to have experienced visions of paradise, "rich fertile country, abounding in game, fish, pleasant hunting grounds and fine corn fields."[42] Gaining the confidence of his people, the prophet preached total rejection of the ways of the children of the evil spirit, from dress and law to property ownership. This was the true way of American life he had acquired in his visions. In the hands of his clever brother, Tenskwatawa would prove to be an effective instrument for generating morale.

Tecumseh's tribal alliance succeeded in winning back natives who had joined the white man's church. Those who resisted were accused of witchcraft and put to death. Future president William Henry Harrison (1773–1841), then governor of the territory of Indiana, appealed to his Delaware allies, many of whom had already deserted, calling Tenskwatawa a false prophet. Why would the Great Sprit choose such an obvious charlatan to be a prophet? Let him prove himself, he told them, by performing a miracle: "If he is really a prophet, ask him to cause the sun to stand still—the moon to alter its course—the rivers cease to flow—or the dead to rise from their graves. If he does these things, you may then believe that he has been sent from God."[43]

Tenskwatawa gave nonbelievers precisely what they wanted. In the first days of June 1806 he assembled his followers in the Indian village of Greenville (now Ohio) and asked them to spread his declared word that he would demonstrate his powers by darkening the sun at midday on June 16. That morning, with tribes assembled, Tenskwatawa heightened the drama by staying out of sight in his lodge. When the sun began to fade into noonday twilight he appeared, shouting to the frightened throng: "Did I not speak the truth? See, the sun is dark!"[44] Then, to the relief of the crowd, the prophet returned the sun to full brilliance. No one in attendance doubted it. Tenskwatawa surely was favored by the Master of Life. His eclipse prediction ignited the spark of the new religion. Under Tecumseh's leadership, it spread to other tribes and prolonged the Indian resistance movement.

Shawnee doctrine soon incorporated tribes from areas of Michigan, Illinois, Wisconsin, and Minnesota. For the next five years the allied natives gained the advantage, until they were finally defeated by Harrison in 1811 at the Battle of Tippecanoe; thereafter the Indians began to lose faith in their prophet. He may have been able to halt the course of the sun, but he could not stop American bullets. Following the inevitable subjugation of the native population, Tenskwatawa, exiled to Canada, died in 1836.

Those who do not believe in miracles (Harrison evidently was not among them) wondered: How did the Shawnee Prophet perform his miraculous feat? During the spring of 1806, scientists had been traveling about the region of Indiana, Kentucky, and Ohio, setting up observing stations to view the forthcoming eclipse. Surely Harrison would have known of these activities—but had he forgotten? Tenskwatawa may have been aware of the astronomers' presence as well. He also knew that Mukutaaweethee Keesohtoa, the Black Sun, was an event that portended war.[45]

Though he did not claim to have predicted it, an African American prophet would use the next American eclipse, an annular one with its distinct ring of fire, visible in Virginia on February 12, 1831, Universal Time, to incite a rebellion of slaves and free blacks on a rampage of plantations in southern Virginia that resulted in sixty white deaths in addition to two hundred of their own. (The eclipse path appears on America's earliest known eclipse map.)

Well acquainted with biblical prophecy, Nat Turner, like Tenskwatawa, was declared a prophet by those who had heard him preach. He claimed to be influenced by visions, among them a message from

First U.S. eclipse map, showing the path of the February 1831 eclipse. (*The American Almanac and Repository of Useful Knowledge for the Year 1831* [Boston: Gray & Bowen], p. 13, from the John Adams Library at the Boston Public Library)

God telling him to take on the yoke Jesus had borne for mankind's sins. He would be the one appointed by God to pick up the fight against Revelation's Antichrist, the white man who had placed his people in servitude. Turner's biographers tell us that in the winter of

1831, God began sending him certain signs in the atmosphere that the rebels should take action. What he interpreted to be the hand of a black man covering the sun appeared on the afternoon of February 12 in what would come to be known as Nat Turner's eclipse. A second (also partial) eclipse followed 177 days (six months) later, on August 13, this time surrounded by a green halo. This was the final signal; eight days later, Nat Turner's rebellion began.

When the early nineteenth-century eclipses took place, tourism did not exist in the United States. It first began to flourish in the Hudson River Valley in the following decade close to the environs of New York City, where wealthy aristocrats acquired easy access to the picturesque Catskills.[46] By 1825, American tourism spread to points farther west, like Niagara Falls, again via the Erie Canal. America was now beginning to shape its own national identity, which meant further distancing itself from European ways (recall John Quincy Adams's appeal to Congress for the creation of the Smithsonian Institution). The scenic American landscape was part of structuring that distinction.

Seeing nature in a unique American setting became one response to the negative forces of industrialization and urbanization affecting the culture.[47] As historian Marguerite Shaffer remarks, the elite classes "turned to the romantic notion of the sublime to enact personal dramas of spiritual transcendent fantasies of the Protestant culture of salvation now threatened by consumerism."[48] By the time of the 1878 eclipse in the Great American West, the subject of the next chapter, touring by rail had begun to flourish. (The Raymond and Whitcomb Tour Company included Pike's Peak on one of its itineraries.)[49] There the gaze of eclipse chaser and eclipse prophet would meet the inquiring eyes of expeditionary forces of astronomers and other scientists pursuing the lunar shadow in search of answers to new questions of their own.

13

Expedition to Pike's Peak, 1878

High on her speculative tower
Stood science waiting for the hour
That darkening his radiant face
Which superstition strove to chase,
Erstwhile with rites impure.

—*William Wordsworth, 1845*

William Wordsworth's eclipse poem draws a sharp line of demarcation between religious omen gathering and scientifically minded data retrieval. Unlike reactions to the 1806 eclipse in America, by century's end there was less room for romanticizing about the wonders of nature. As author Theodore Dwight wrote at the time: "Instead of wishing to see the world through a fancied medium, the rational traveler wishes to view it as it is."[1] Romantics, he believed, didn't derive knowledge based on disciplined study; emotion got too much in the way.[2] Historian Peter Gay argues that the period's new skepticism toward religion developed out of a recovery of nerve that the scientific understanding of things could lead to a better world. Here the still ongoing war between science and religion began in earnest.[3]

The Victorian age in Great Britain was a time when adventurous professionals—zoologists, geologists, and Arctic explorers—received both government and private support to launch expeditions to far-flung regions of the globe in search of answers to their biggest questions: What makes the sun shine? Why is the corona so hot? Is there a phantom planet yet to be discovered in the solar system? Members of the Royal Astronomical Society organized an Eclipse Committee, funded by the U.K. Treasury, as eclipses began to shed the image of cosmic phenomena to be gawked at in trepidation and instead became the focus of serious scientific attention, worth studying *as they are.*[4]

Anticipating two eclipses due to take place within their continental borders in 1869 and 1878, and eager to demonstrate scientific independence from Europe, American astronomers were primed to organize their own expeditions. Earlier American group eclipse trips had been local events sponsored by college observatories and led by resident astronomers.[5] The press paid due attention: in the 1870s the number of words per page in the *New York Times* devoted to astronomy increased tenfold over the previous decade.[6] Expeditions tied to the August 7, 1869, solar eclipse included British astronomers, but they were organized largely by their American colleagues. With the cooperation of the nascent railway system, professional sun watchers set up observing stations along the path of totality from the Dakotas to the Carolinas.[7] Participating British astronomer Sir Norman Lockyer was impressed with the way "the government, the railway, and other companies and private persons threw themselves into the work with marvelous earnestness and skill. . . . There seems to have been scarcely a town of any considerable magnitude along the entire line, which was not garrisoned by observers, having some special astronomical problem in view."[8] It is worth noting that at the next eclipse, in 1870, in Cadiz, Spain, the British reciprocated by inviting American astronomer Simon Newcomb to join their expedition. They even paid a portion of his expenses.

Lunar shadow chasers were still puzzling over the nature of the corona. They would document it photographically for the first time, no mean task considering the woeful state of the art of high-speed photography in America at the time.

Spectroscopy was another new technology undergoing rapid development. Even poet Ralph Waldo Emerson wrote about it: "Who would live in the stone age or the bronze or the iron or the lacustrine? Who does not prefer the age of steel, of gold, of coal, petroleum, cotton, steam, electricity, and the spectroscope?"[9] By the early 1800s, astronomers had discovered that when sunlight was passed through a narrow slit in front of a prism, the resulting band of rainbow colors was marked by hundreds of thin black lines. They could read the chemical signatures of the lines by matching their wavelengths alongside spectra produced by known sources in a laboratory. The black lines represented portions of the continuous colors emitted by radiation welling up out of the hot solar interior, which had been partially absorbed by the cooler gases lying on top. Thus they identified atoms of hydrogen, iron, and other elements as constituents of the sun's lower atmosphere. Another of the elements identified (in 1868) was helium, discovered in the solar spectrum prior to having been known on earth. When astronomers turned their spectroscopes on the corona, they found that it radiated bright lines instead, an emission spectrum, which usually implies high temperature. Here was a layer of the sun's atmosphere hotter than the photosphere below. One unidentifiable bright line in the green part of the spectrum was a major puzzle. Astronomers were unable to detect its counterpart in earthbound laboratory comparisons. They called it "coronium," a mystery element present only in the sun.

What made the corona shine? When sixteenth-century astronomers had first observed it, they thought its effects were produced by the sun backlighting and filtering through the moon's atmosphere, until the instantaneous eclipse of stars by the moon proved the moon has no atmosphere. Could the corona then be the result of light scattered by the earth's atmosphere? Bolstered by observations of the intense Leonid meteor storms observed in 1833 and 1866, another theory argued that swarms of meteors in orbits passing close to the sun contributed a large infall of matter that ignited, causing the illumination. Or could the sun's crown be part of the zodiacal light, a luminous band produced by reflection of sunlight off particles in the plane about which the planets orbit the sun?

By 1870, evidence began to weigh heavily in favor of an internal explanation of the corona, especially when it was realized that its shape varied with the activity of the eleven-year sunspot cycle, and that coronal streamers, those streaky rays extending outward from the eclipsed sun, appeared to follow the sun's magnetic force lines. The visible red flames, or prominences, between the corona and the photosphere further suggested an internal energy source.

On the eve of the great American eclipse of 1878, there were still more questions astronomers needed to put to the test: How far out does the corona extend? What causes it to shine in such peculiar colors? Why the streamers, and why does their placement shift so radically, from a nearly even distribution around the disk during sunspot maxima in 1870 and 1871, to a pattern running out several radii at the equator but clipped off at the pole, the way they looked in 1874 and 1875? Were the two kinds of rays connected? From this point on, American-led scientific investigations of the sun—and especially its corona—would hold center stage on eclipse expeditions.

Today's astronomy is all about probing the outer limits of a much more vast universe, finding the most distant galaxy, discovering

gravitational waves, investigating dark matter, observing planets orbiting faraway stars, and searching for Pluto-like objects in the outer
reaches of the solar system. Nineteenth-century skywatchers seemed
more concerned with pointing their scopes inward. Their passion was
fueled by a different kind of Planet X fever: searching for planets inside the earth's orbit.

"It is not improbable that the detection of Vulcan may be merely
the first in a series of similar discoveries. . . . The prospect of planetary
discoveries in this part of the system is at present more hopeful than
in the space beyond the orbit of Neptune," wrote Indiana University
astronomer Daniel Kirkwood in 1878.[10] Aptly named for its hellish
locale, the intra-Mercurial planet (thought to lie between the sun and
the orbit of Mercury) had been the object of intensive skywatching in
the sun's vicinity, especially during solar eclipses, when much of the
sun's glare is eliminated.

Long before *Star Trek*'s Mr. Spock made the name famous, Vulcan's hypothetical existence was deduced by the brilliant French
mathematician-astronomer Urbain Le Verrier. Born of modest means
(his father was a low-level bureaucrat), Le Verrier was a gifted math
genius despite having secured only a provincial education; his ambi-
tion and innate skill, however, together with his father's decision to
sell his home in order to finance the boy's education, got him admitted to Paris's prestigious École Polytechnique, from which he gradu-
ated with honors. There he also acquired his first appointment in
astronomy and set out to study the stability of the orbits of planets, a
hot topic in nineteenth-century astronomy. Le Verrier's claim to fame
is his prediction of the discovery of the planet Neptune. He calculated
the position of the trans-Uranian planet (then the farthest known
planet from the sun) based on tiny disturbances observed in its orbit.
Writing down the precise coordinates, he confidently predicted that if

astronomers pointed their telescopes at that location, they would find it. The planet Neptune was discovered in 1846, the same year the brilliant Frenchman predicted its exact whereabouts. "I have already begun to mount the ladder of success," he immodestly wrote in his diary, so "why shouldn't I continue to climb?"[11]

A decade earlier, Le Verrier had already turned his attention to tiny deviations detected in the orbit of Mercury. In the course of making one eighty-eight-day revolution on its elliptical orbit, the planet was not returning to its original place, but rather advancing about 1¼ seconds (one-sixtieth of one-sixtieth of a degree) per revolution. In other words, Mercury's orbit was open-ended, precessing like a toy top. Le Verrier calculated that the so-called advance of perihelion (the point on the orbit nearest the sun) amounted to a minuscule thirty-eight seconds per century. Following Newtonian theory, he thought the disturbance could only be caused by a missing mass following an orbit located somewhere in the space between Mercury and the sun, exerting its gravitational pull, thus tugging Mercury out of its orbit. The math was more complex than the Neptune calculations, but Le Verrier managed, with pen and ink, to generate predictions that excited astronomers because they could be tested by observation. In the middle decades of the nineteenth century, astronomers began directing their tubes toward the solar vicinity. They repeatedly thought they discovered the planet, only to lose, rediscover, and lose it again. Most Vulcan sightings turned out to be sunspots, dark, cooler blotches of gas in the solar photosphere projected against the brighter background. So, where was the elusive intra-Mercurial planet? This question drew increasing attention among late nineteenth-century American astronomers. What they needed was an eclipse that would offer them the best conditions to accommodate their newly developed technology, an eclipse they could witness at high altitude, where the air is thin and clear.

On July 29, 1878, that need would be fulfilled. The moon's shadow was due to drop down in northeast Asia, cross the Bering Strait, and enter U.S. territory in newly acquired Alaska. It would then glide down the Aleutian Islands, move across Canada, and enter Montana and Wyoming. Totality would happen in the 116-mile-wide strip passing through the Rocky Mountains of Yellowstone, the Medicine Bow, and over the 14,000-foot summit of Pike's Peak, before heading on to the Indian Territory of Oklahoma and the Texas-Louisiana coast, then out into the Gulf of Mexico. Despite Indian problems deterring some eclipse chasers from southern Wyoming,

Thanks to the clear thin air, viewing conditions for the Rocky Mountain eclipse of 1878 were spectacular enough to catapult it to the cover of *Harper's Weekly*, August 24, 1878. Long streamers issue from the corona at sunspot minimum.

that area nonetheless emerged as a favorable destination because the Union Pacific Railroad was under construction there. The Colorado Rockies were another desirable location. The Denver area attracted a lot of attention from the press for at least a year leading up to "E-Day": "The world has never witnessed such extensive preparations," heralded the *Laramie Daily Sentinel.* "Scientists from every quarter of the globe inhabited by intellectual beings are today within the limits of totality, awaiting with prayerful interest the grand moment."[12]

Government support for the scientific enterprise of 1878 was a first on the west side of the Atlantic. Congress granted the U.S. Naval Observatory $8,000 to set up eight observing stations along the eclipse path, ranging from Virginia City, Montana, to the Texas hill country. Weeks before the eclipse, a cadre of astronomers took up residence in solitary hotels in Central City, Colorado, and Creston and Rawlins, Wyoming. The *Boulder County News* dubbed the visiting sages "the wise men from the east . . . the most modest and unpretending, and unassuming, and unostentatious anywhere to be found."[13]

Thomas Edison went west, too, to Rawlins. At the invitation of wealthy physician/amateur astronomer Henry Draper, the celebrated inventor proposed to unveil his palm-sized gadget—an ultrasensitive heat measuring device he called a tasimeter. He boasted that it could detect the heat of a lighted cigar in the mouth of a man entering the far side of a room.[14] If the corona was tied to the sun, it ought to be hot enough to emit infrared radiation, and the ever boastful Edison promised that his apparatus would settle the question once and for all.

Only thirty-one years old at the time, Edison already held eighty-nine patents in telegraphy alone, including the stock ticker.

Newspapers loved him, and so did the railroad, offering him free passage to the site. "Professor Edison attended by a party of scientists," announced one headline.[15] Even though he got top billing over the astronomers, Edison couldn't resist expressing his distaste of professional academics: "I wouldn't give a penny for the ordinary college graduate, except those from Institutes of Technology . . . [who] aren't filled up with Latin, philosophy, and all that ninny stuff."[16] He derided the astronomers' worksheets, remarking that they resembled the timetable of a Chinese railroad.

Everyone in town went out of their way to get a look at the great inventor. Edison later told this story of his encounter with a cowboy, shortly after checking into his shared room at the only hotel in town: "After we retired and were asleep a thundering knock on the door awakened us. Upon opening the door, a tall, handsome man with flowing hair, dressed in Western style, entered the room. His eyes were bloodshot and he was somewhat inebriated. He introduced himself as 'Texas Jack' . . . and he said he wanted to see Edison as he had read about me in the newspapers."[17] During the conversation, the drunken intruder reportedly proceeded to demonstrate his gun-handling skills by shooting the weathervane off the top of the train station across the street. Edison eventually cajoled him into leaving.

The most obsessive of all the Vulcan questers also set up shop in Rawlins. Born in 1838 a poor Ohio farmboy, James Craig Watson started ed working in a factory at the age of nine. When it closed, he switched to selling books and apples at the Ann Arbor rail station and later secured a job as a machinist. All the while he learned Latin and Greek on his own and at the age of fifteen was admitted to the free-tuition University of Michigan. There, Watson came under the tutelage of Franz Brunnöw, director of the Detroit observatory, who would come to regard young James as his "only student." Possessed of as much self-confidence as his hero Urbain Le Verrier, he once entered his name in one of his notebooks as "James Watson, the Astronomer Royal." Excelling in telescope making and rapid-fire computing (usually in his head), Watson quickly matriculated from observatory assistant to professor of astronomy and observatory director, replacing his mentor when he left for another job. Once established, Watson turned his attention to the fashionable enterprise of chasing asteroids.

Since the unprecedented discovery of minor planet Ceres in 1801, followed a few years later by Vesta, Pallas, and Juno, astronomers were surprised that the space between the orbits of Mars and Jupiter was populated not by a single large planet, but instead by a band of minor planets. Discovery necessitated recognizing the so-called asteroids as moving objects against the background of distant stars. Watson mathematically charted the position of every fixed star lying along the annual path of the sun among them, so that intruders might stand out, especially during eclipses. But in the competitive world of asteroid hunting, his efforts would not go unrivaled.

When I accepted my position as Colgate's only astronomer, I was surprised to learn that our neighboring institution, Hamilton College, which offered no astronomy courses at the time, once housed one of America's largest telescopes, and that the now little known astronomer who set up the thirteen-inch refracting instrument at the Litchfield Observatory in 1858 had made some of the most important astronomical discoveries of that century. Christian ("C.H.F.") Peters, a prickly character with eclectic interests, had already traveled the world, from Sicily, where he studied volcanoes and discovered a comet, to Constantinople, where he became fluent in Turkish and Arabic. He served as scientific adviser in the court of the Grand Vizier to the Sultan, and there acquired the use of a handsome eleven-inch refracting telescope, which he used to pursue his asteroid-hunting predilections. He quit in a huff after a dispute with the sultan's astrologers over how to interpret celestial events, and ultimately landed penniless at the tiny college in rural Clinton, New York. There he set up his own court and engaged in a heated competition with Michigan's Watson over who would chart the most asteroids. Being a generation older than Watson, Peters had a sizable leg up (C.H.F. would emerge victorious in the contest by a count of 48–22). He was never a fan of anyone else's claimed intra-Mercurial planets, proclaiming, "I want to see them myself!"[18]

Watson had already failed to detect the elusive Vulcan on two previous expeditions, but, despite his earlier failings, he radiated optimism. An altitude of six thousand feet accorded him the decided advantage of operating above the dust and humidity of the lower part of the earth's atmosphere. Now he would have his all-time best shot at detecting the mystery planet. In addition to his piano, Watson brought along a four-inch refracting telescope, which he planned to use in executing sweeps across the area immediately adjacent to the eclipsed sun "to search for any planetary bodies which might be visible in the neighborhood of the sun."[19] He would be allowed only two minutes and fifty-six seconds to complete the task. Watson had a plan in place should he spot any interloper in the star field. He would telegraph a message from the adjacent Union Pacific rail station to the one in Dallas. There he would hire a man on horseback to personally carry the coordinates in hand, racing the shadow a quarter-mile to another astronomical observing station, where the discovery could be confirmed moments later when totality arrived. The race between telegraph and lunar shadow would be an eclipse first.

A Smithsonian-funded project at the top of 14,000-foot Pike's Peak resulted in the most ambitious of American eclipse expeditions to date. The major contributor was astronomer, physicist, mathematician, inventor, and aviation pioneer Samuel Pierpont Langley. In addition to founding the Smithsonian Astrophysical Observatory (in 1890) and directing Pittsburgh's Allegheny Observatory, Langley had also helped pioneer early air travel. In 1887 he began experimenting with rubber-band-powered models of heavier-than-air craft and gliders, before moving on to building steam-powered flying machines. In 1896 he flew an unpiloted plane three-quarters of a mile, more than ten times the distance achieved by any previous craft. He even secured a grant from the War Department to build a piloted plane, but before he succeeded, the Wright brothers came along (in 1902). He gave up the effort after two crashes in 1903. Writer/poet Rudyard Kipling wrote of his meeting Langley at the time: "Through [Theodore] Roosevelt I met Professor Langley of the Smithsonian, an old man who had designed a model aeroplane driven in a marvel of delicate craftsmanship. It flew on trial over two hundred yards, and drowned itself in the waters of the Potomac, which was a cause of great mirth and humour to the Press of his country. Langley took it coolly enough and said to me that, though he would never live till then, I should see the aeroplane established."[20]

As professor of astronomy at the Western University of Pennsylvania (today the University of Pittsburgh), Langley worked in solar research and spent quite a bit of time on the invention of a bolometer, a device like Edison's tasimeter, designed to measure infrared radiation. Financing eclipse expeditions was a major problem; so, to raise money for his observatory, the resourceful Langley offered (for a modest fee) to provide accurate time signals to the railroads. In the newly developed high-speed world of rail travel, train operators needed to keep the same time as switch operators and station masters, which made the service Langley offered a valued commodity.

The ambitious eclipse agenda Langley set up on Pike's Peak was multifaceted, and he turned to his talented brother John, a University of Michigan chemistry professor, to assist him. Langley focused his attention on collecting as much data as possible about the corona. To that end, he brought along a special thirty-eight-foot-long, five-inch-diameter solar telescope in addition to another "portable" unit weighing over a thousand pounds. But first, rain-damaged boxes of equipment arriving at the station awaited unpacking and repacking before the trip up. Always resourceful, Langley raided the cupboard for lard to smear on the connecting parts of his scopes to keep them from rusting.

Langley hired porters and assembled a mule team as he initiated the eighteen-mile climb to the summit; there was no road. Some of his supplies had to be abandoned on the way up because porters were unwilling to take on the excess burden. Cleveland Abbé, the Cincinnati Observatory astronomer and one of the leaders of the expedition, was plagued with altitude sickness on the way up and needed to be taken down the mountain on a stretcher. When Langley finally arrived at the summit of Pike's Peak, three days before his camping supplies got there, he found the terrain so uneven and boulder-strewn that he could scarcely clear a space big enough to pitch a tent. Each instrument required support piers that needed to be constructed out of whatever stone and wood could be found on site. Undaunted, his inventive brother improvised support piers for the telescope by fashioning spikes out of ambient wood and driving them into the ground between boulders, stabilizing them to create a wooden frame. He filled the space with rocks and nailed atop it a platform made of planks to make a firm surface for the delicate instruments.

Weather is the biggest bugaboo of all eclipse expeditions, and the forecast, to judge from meteorological conditions throughout the

Rockies in the two weeks leading up to July 29, was anything but auspicious. Torrential downpours flooded the valleys the day before the eclipse, and Langley's team emerged from half-blown tents only to be greeted by a ten-inch blanket of snow. The Sunday prior, Denver churches had conducted special services, including prayers for good weather.

Evidently they worked. From Cheyenne to Pike's Peak, eclipse day dawned sparkling clear. "Whenever the astronomers want an eclipse that is an eclipse, let them arrange for it in Colorado," the *Denver Daily Times* reported. The *Colorado Banner* added: "This state has furnished the grandest eclipse of the age."[21] One observer exclaimed that the spectacle was "the grandest sight I ever beheld. . . . It scared the Indians badly. Some of them threw themselves upon their knees and invoked the Divine blessing; others flung themselves flat on the ground, face downward; others cried and yelled in frantic excitement and terror."[22] Scientific responses to eclipses on many expeditions have often been held aloft the superstitious behavior of the local population. A derisive Denverite remarked, "Even John Chinaman was expecting the eclipse," but rather than sketching the corona, "the almond-eyed celestials . . . beat their gongs through totality" to chase away the dangers.[23] Historian Alex Soojung-Kim Pang, who recounts many such eclipse stories told during the Victorian era, thinks they say more about the anxieties of Westerners than beliefs of the natives.

The scientific investigations proceeded. At 3:20 in the afternoon the darkness rolled out of the west and swept over Pike's Peak; then the last tiny sliver of sunlight vanished. Langley had his equipment set up to measure the brightness, or candle power, of the corona: a candle mounted on a track inside the thirty-four-foot-long telescope was slowly drawn by a long string until its brightness was judged by eye to be equal to that of the corona when both images were projected onto a paper screen. Data from a group of volunteers who sketched the corona showed it extending outward from the disk to an unheard-of distance of ten diameters, five million miles. It had an extremely elongated shape, typical of what had previously been seen at solar minimum.

No question, Langley concluded, the corona was surely a part of the sun; but he was disappointed because his infrared-detecting bolometer failed him. He was unable to detect the heat of the corona with the relatively insensitive device.[24] The 1878 eclipse also marked the first clear visible spectroscopic recognition of the reversal of the sun's dark absorption lines (from the photosphere) to bright emission lines (from the chromosphere and corona) at the onset of totality. The bright-line spectrum would not be photographed until 1896. (Today, astronomers no longer need a total eclipse of the sun to view the sun's corona. They can do it with a coronagraph, a telescope attachment featuring a built-in disk that takes the place of the moon to block out the light of the photosphere. Originally invented in 1931 by French astronomer Bernard Lyot, modern versions of the device, used in the Hubble Space Telescope, are capable of revealing coronae around other stars, and they are frequently employed in the search for extra-solar planets. Space-based observatories also enable access to the outer corona out of eclipse. Because the earth's atmosphere scatters sunlight rendering coronal observations outside one solar radius from the sun's radius inaccessible, astronomers have taken to equipping space telescopes with coronagraphs that make streamers accessible out to more than thirty radii.)

Moments earlier at Rawlins, Edison, never one to spend much time rehearsing, was setting up his tasimeter in the only available protective enclosure from the wind in the astronomers' camp, a chicken coop, "in a small yard enclosed by a board fence six feet high; at one end, there was a house for hens. I noticed they all went to roost just before totality. At the same time a slight wind arose and at the moment of totality the atmosphere was filled with thistle down and other light articles."[25] Over the years, the "chickens-came-home-to-roost" story became embellished in a way that suggests the birds wrecked his experiment. Edison's infrared-detection instrument proved to be a good example of precisional overkill. The instant he pointed the telescope containing it at the corona, the needle got pinned. Still, the result was positive. Though Edison didn't get any numbers, he immediately concluded: "the sun's outer shroud did indeed give off infrared radiation."[26] A *New York Herald* reporter offered this account: "When but one minute of totality remained Edison succeeded in crowding the light from the corona upon the small opening of the tasimeter. Instantly, the galvanometer cleared its boundaries. Edison was overjoyed."[27]

Nearby, young Watson, who already had memorized the position and brightness of every fixed star in the area, began his sweeping search just after 3:13 p.m. in three-sun-diameter-wide bands to the east and west of the totally eclipsed disk. He carefully marked the position of every detectable object he saw on the properly oriented cardboard-mounted paper circles on which he projected his images. As the three-minute clock ran down, nothing unusual was sighted; then, seconds before the end of totality, Watson spotted, close to the star Theta in the constellation of Cancer, "a ruddy star whose magnitude I estimate to be 4½."[28] Excitedly, and without bothering to wait for confirmation, he announced his result to a reporter standing by from the local Laramie *Daily Sentinel,* who wrote: "Professor Watson, of Ann Arbor, Michigan, who is now the most noted astronomical observer and discoverer in the world, had taken the job of FINDING VULCAN. And he found it." Hours later the *London Times* broadcast the story to the world; the *New York Times* enthusiastically chimed in: "The planet Vulcan after so long eluding the hunters . . . appears at last to have been fairly run down and captured."[29]

Like many scientific findings reported prematurely, the discovery of the new planet did not withstand the test of time. The problem began when Watson discovered an error in his clock. Correcting his calculation of the mystery planet's position, he concluded that what he thought was Vulcan landed too close for comfort to the location of a known star. To make matters worse, no other astronomer could verify Watson's sighting. He was later forced peevishly to walk back his discovery, stating that although he couldn't positively conclude that what he saw was Vulcan, "I certainly have the right to express my honest belief that [it is]."[30] Watson's nemesis quickly jumped into the fray. C.H.F. wrote a scathing review of Watson's Pike's Peak efforts in one of the professional journals, complaining that Watson's observations were inaccurate, gathered hurriedly in too dim a light, and otherwise flawed: "It is . . . apparent, to every unbiased mind, that Watson observed [stars] and nothing else."[31]

Watson vowed that in some future eclipse he would regain the opportunity to give solid reasons to support his faith, but his early death two years later, at the age of forty-two, would prevent him. What one historian of astronomy termed the "recurrent nightmare of Vulcan" would continue with spurious sightings, though with greatly diminished frequency, all the way into the next century.[32] In the draft of his report to the Smithsonian on an eclipse expedition he led twenty-two years after the Pike's Peak event, the venerable Samuel Pierpont Langley commented, "It was further learned that Professor Pickering . . . desired; if possible, that a search [for Vulcan] with similar apparatus should be made by other eclipse parties. This observation was with some hesitancy added to the programme, for I had no strong expectation that anything but negative evidence would be secured."[33]

If not Vulcan, what was the missing mass responsible for Mercury's aberrant motion? Was there more than a single tiny planet or perhaps an undetectable ring of asteroids within Mercury's orbit? The answer, quite simply, was that there *was* no missing mass. Working like Le Verrier, with pen and paper, in 1915 Albert Einstein concluded that the advance of Mercury's perihelion resulted from what would become known as his general theory of relativity. Specifically, the theory explained that massive bodies like the sun warp the space around them, and this warping was responsible for deviating Mercury's motion from its anticipated course. Einstein calculated that Mercury's excess motion should amount to forty-three arc seconds per century, exactly what astronomer Simon Newcomb, correcting Le Verrier's earlier thirty-eight seconds, had concluded from telescopic observations of the planet.

The definitive verification of the theory came during the solar eclipse of May 29, 1919. If massive bodies like the sun warped space, then that warping would influence not only the motions of planets, but also the path of light itself. During the eclipse, British astronomer Sir Arthur Eddington recorded the locations of stars in a cluster in Taurus known as the Hyades, which stood adjacent to the eclipsed disk. At dual stations along the shadow path in northern Brazil and West Africa, astronomers took photographs of the cluster members and compared them with those obtained when the Hyades were at a different location in the sky earlier in the year. Only in the darkness of totality was it possible to record stellar positions so close to the bright (and massive) sun. Comparing the locations, they verified the bending of the stars' light and confirmed Einstein's predicted effect, called gravitational lensing. And therefore, the 1919 event was widely proclaimed *the eclipse that changed the universe.*

As for Vulcan, and the mysterious element coronium, these were mere illusions, ghosts exorcised by the progress of science.[34] Gradually, the study of astrophysics, which deals with the inner workings of stars, displaced celestial mechanics, or the study of stars' movements, on the astronomers' agenda. New direct observation spectroscopes and photographic spectroscopes (spectrographs) permitted analysis of not only the chemical composition of the sun, but also how temperature and pressure in the sun's atmosphere affect the formation of spectral lines. In the 1930s, astronomers found that the "coronium" spectral signature was actually produced in part by iron atoms with thirteen electrons removed by the extraordinarily high coronal temperature. Coronium was not the only "element" to suffer this fate. "Nebulium," so named because it was believed to produce a bright green line in the spectrum of the Orion Nebula, turned out to be nothing more than oxygen, also missing electrons. Astronomers were beginning to realize that the entire universe is made up of pretty much the same stuff—ambient conditions disguise what that stuff looks like. As Copernicus wrote four centuries earlier: "We thus follow Nature, who producing nothing vain or superfluous often prefers to endow one cause with many effects."[35]

Maria Mitchell (1818–1889), widely celebrated in the scientific community as America's first woman astronomer, also traveled west to witness the 1878 event. She was a Nantucket native, one of ten children born of Quaker parents who believed in educating their girls as fully as their boys. She developed a passion for astronomy at an early age. Taught by her father, she learned to sweep the sky every clear night with his brass telescope, perched on the roof of their house. She discovered a comet in her twenties and became the first woman elected to the American Academy of Arts and Sciences and the American Association for the Advancement of Science. She also became the first woman appointed to the faculty of Vassar College, in 1865, as professor of astronomy and director of the observatory. When she learned her salary was less than that of many younger male professors, she demanded—and received—a salary increase.[36]

Professor Mitchell and a handful of her Vassar College students, some never having been away from home, boarded a train in Boston and traveled to Denver via Cincinnati, Kansas City, and Santa Fe. Unfortunately, only one of her telescopes, minus its lens, made it all the way, thanks to a war between competing railroads. Mitchell and her girls haunted the telegraph room attempting to track the whereabouts of the equipment; it finally arrived just three days before totality. The weather on the outskirts of the capital was no better than what Langley's group was experiencing high on Pike's Peak: Friday, occasional rain; Saturday, steady rain; Sunday, heavy rain mixed with hail. But Monday, eclipse day, dawned bright and clear at the elevated plateau on the outskirts of the city where the conclave had managed to set up their tents and instruments. Mitchell assembled her young charges and offered them these inspiring instructions:

You will see nature as you never saw it before—it will neither be day nor night—open your senses to all the revelations. . . . Let your eyes take note of the colors of Earth and Sky. Observe the tint of the sun. Look for a gleam of light in the horizon. Notice the color of the foliage. Use another sense—notice if flowers give forth the odors of evening. Listen if the animals show signs of fear—if the dog barks—if the owl shrieks—if the birds cease to sing—if the bee ceases its hum—if the butterfly stops its flight—it is said that even the ant pauses with its burden and no longer gives the lesson to the sluggard.[37]

Minutes before first contact, the professor imposed silence, then the wait: "for even time is relative and the minute of suspense is longer than the hour of satisfaction."[38] Following two minutes and forty seconds of mostly naked-eye observation, Mitchell, echoing comments made by Warren de la Rue, wrote: "And now we looked around. What a strange orange light there was in the north-east! What a spectral hue to the whole landscape! Was it really the same old earth, and not another planet? . . . Great is the self-denial of those who follow science. They who look through telescopes at the time of a total eclipse are martyrs; they severely deny themselves. The persons who can say that they have seen a total eclipse of the sun are those who rely upon their eyes."[39]

Turn-of-the-century eclipse chasing loomed as an exciting career option for budding astronomers. With nearly two dozen eclipses due to occur across the globe in the four decades centered on 1900, and with safer, speedier modes of travel and improved detection methods, young David Todd (1855–1939), newly minted master's degree in hand from Amherst College, also chased the shadow in 1878. He chose to go to Texas for the first of his twelve expeditions, most of which were planned when he later returned to Amherst as professor of astronomy and director of the observatory, which he built. Sadly, Todd's encounters with totality would spin a tale of undeserved woe, especially given his considerable engineering computational and astronomical skills. Originally he wanted to be an organist, until he became more curious about the internal workings of the instrument than how to play a tune on it.

Like his European predecessor by a generation, Warren de la Rue, Todd was a genius at mechanical design, especially when it came to transporting and setting up cumbersome, complex eclipse-monitoring equipment. He developed a forty-foot focal-length telescope for photographing the corona. To eliminate nervous tension in eclipse photography, he invented an automatic contraption that allowed him to take hundreds of exposures during just a few minutes of totality. It consisted of a series of levers, cords, and pulleys that enabled him to operate twenty-three telescopes simultaneously, all of them aligned perfectly parallel and pointed directly at the solar target. The apparatus included mounting, exposing, and developing large, sensitive glass plates.[40] The earliest form of Todd's device inserted the plates into position at the focus of each instrument, exposed them, then removed and replaced them. It was operated via pneumatic power using a foot pedal, just like an organ. Properly placed perforations on a cylinder issued the commands.

Mechanical ingenuity and careful record-keeping usually go hand in hand, but Todd took the latter to extremes. In one of his copious

Astronomer David Todd's pneumatically operated photographic plate dispenser for telescopic eclipse observations. Note the organ pedals. (From David Todd's *Stars and Telescopes* [Boston: Little, Brown, 1901])

THE PNEUMATIC COMMUTATOR AND PHOTOGRAPHIC BATTERY OF
ECLIPSE INSTRUMENTS (TODD)
(As mounted at Cape Ledo, Africa, for the total eclipse of December 1889)

diaries, he confesses that he should have been an engineer rather than an astronomer. Included among other detailed diary entries were the number of hours and minutes spent at the telescope, the number of hours he slept (he averaged two to three hours a day), and the phase of the moon when he engaged in any activity.[41] His wife, Mabel Loomis Todd (1856–1932), wrote heartfelt passages about what it was like to witness an eclipse and in addition wrote a book about eclipses. (She also edited several books that brought some of the poet Emily Dickinson's most influential work to light, all the while carrying out a scandalous affair with Austin Dickinson, brother of the poet.)[42]

As a teacher of astronomy, I've long been aware of the proliferation of women who wrote about astronomy, especially eclipse chasing, in the early twentieth century. This was a time when women's education was undergoing distinct advances. Rebecca Joslin was a Smith College student who traveled with her teacher to Spain to view the August 30, 1905, eclipse, then again in 1914 to war-torn Europe for the August 21, 1914, event. She lost out to clouds on both attempts before catching the big one in New York City in 1925. Isabel Martin Lewis, who wrote a handbook on solar eclipses, was formally trained in astronomy at Cornell University, later securing a job as a "computer" at the United States Naval Observatory. There she met and wed Clifford Lewis, a much higher-salaried professional stargazer.

Women's work in astronomy then consisted of organizing and computing astronomical data; they were hired as *calculators*. The select group who worked on photographic plates of star fields and stellar spectra at Harvard under Edward Pickering's direction became known as "Pickering's Harem." More than a dozen women worked in a large room on the top floor of the Harvard Observatory. The men, except for Pickering, who occupied an office across the hall, all labored on the lower floors.

David Todd spent more than a year carefully planning for the expedition to Yokohama, Japan, to document the August 19, 1887, eclipse. He constructed specially modified shutters and plate holders and a light-proof camera to fit the telescope. He also hired an experienced Japanese photographer to assist him as they repeated the drill several times: partial phase photos every fifteen seconds, then at totality, exposures from fifteen down to one second, minimizing the loss of time at five seconds in between. "With so efficient a photographic corps, and the drill which we all underwent, I had the best of reasons for anticipating complete success," he wrote.[43] After a month spent hauling equipment across the continental United States to the west coast, before taking a steamship to Yokohama to search the country for over a month to acquire a suitable site, Todd made his choice. Unfortunately, a nearby volcano erupted and belched smoke the night before the event. The next morning the sky dawned clear, but moments before totality the clouds moved in, and they stayed there until the eclipse ended. Foiled again, David said little, but the day after, Mabel wrote (to Austin):

> Oh! my dear, my dear! How can I find words to tell you of yesterday's disappointment! The sky was as clear as I ever saw it . . . and when I went up by the instruments about one o'clock, I saw the least speck of a white cloud near the western mountains. It rose with immense rapidity and covered most of the western sky within an hour. . . . I can't write about it, for I am too heart-sore. . . . Well, the chief of the party [her husband] stood it like a hero and philosopher, I knew that he was losing what up to this point is the chance of his life, for his method of observing this [recording an eclipse by photoheliograph] had never been used at an eclipse before and all

astronomers were watching—he had worked day and night, oblivious to sickening heat, or working all night long, and everything was in perfectly brilliant condition for a world-wide success. . . . I would have given anything (but our one joy), out of my own life, to have cleared away those clouds for only three minutes—just the totality. . . . I cannot express my pain at the terrible afternoon yesterday.[44]

David had bad luck again in Japan in 1896. "Only five more working days before the day that must bring us the corona or bitter disappointment," wrote Mabel, who accompanied him.[45] At first contact the sun was partly obscured by a cloud mass, which thinned out to reveal a corona only dimly seen: "Well might it have been a prelude to the shriveling and disappearance of the whole world—weird to horror, and beautiful to heartbreak, heaven and hell in the same sky."[46] Just a few pictures of the blurred corona resulted. Only after many setbacks did David finally strike gold with his automaton during the August 30, 1905, eclipse in Tripoli, where he obtained several hundred exposures.

Watson, Langley, Mitchell, David Todd—all were remarkable in their capacity to surmount the rigors of nineteenth-century travel, carting their cumbersome instruments to faraway places for brief encounters with the shadow. Even if favored by the skies, astronomer and historian of astronomy Donald Osterbrock reminds us, "the tension of eclipse observations in which delicate scientific equipment must work at a particular moment of time completely outside the observer's control . . . is unimaginable."[47]

14

New York's Central Park, 1925

The sun may be eclipsed, but New York—never!

—*Mayor John ("Red Mike") Hylan, 1925*

Noted Dutch astronomer and father of planetary science Gerard Kuiper was my teacher.[1] Kuiper, perhaps best known for his eponymous Kuiper Belt of minor planets (now including Pluto) beyond the orbit of Neptune, was way ahead of his time. I recall the day he excitedly entered the classroom with a hand-drawn graph. Based on his spectroscopic studies of the earth's atmosphere, the plot displayed the percentage of carbon dioxide in the atmosphere over time. It included data going back two generations before the invention of the internal-combustion engine. He pointed to the ever-steepening upward trend, especially in the two most recent decades, in the line he'd drawn to connect the dots. Then, in a thick accent, Kuiper made the first reference I'd ever heard to the greenhouse gases and the problem of global warming (this was 1960).

Kuiper was one among a host of brilliant Dutch astronomers who dominated the discipline in the mid-twentieth century. They came out of the Leiden school established in the early decades by Ejnar Hertzsprung. My teacher often spoke with pride about the accomplishments of his countrymen-peers: Jan Oort, who mapped our off-center locale in the Milky Way Galaxy; Maarten Schmidt, discoverer of the pulsing quasars; and Willem Luyten, Hertzsprung's first doctoral student, who charted the sun's path through the Galaxy by measuring the motions of 400,000 stars in the vicinity of the solar system. Luyten would also become the eye (singular) of America's most watched solar eclipse ever.

In 1925, one out of sixteen Americans lived in New York City, which had just surpassed London as the world's largest city. It also boasted the world's tallest structure: the 792-foot Woolworth building, and the most up-to-date water, sewage, and electrical systems. With low unemployment, and per capita income experiencing a healthy increase during the postwar era, New York's population had more time for leisure than previous generations. Theatergoers paid up to $3.50 a seat to help pack one of the fifty new musicals to open in a single season of live theater, including Vincent Youmans's *No, No, Nanette,* Rodgers and Hart's *Dearest Enemy,* and Kern and Hammerstein's *Sunny.* This was Broadway's golden age. Two of George Gershwin's symphonic jazz pieces, *Rhapsody in Blue* and *Concerto in F,* debuted, and *Ben-Hur* (two years short of sound flicks) was drawing large audiences at motion picture theaters.

There were problems in New York too. The city was nearly bursting at the seams of its poor infrastructure. There were no tunnels, no George Washington or Triborough Bridges, and no West Side Highway. Complaints about traffic jams (there were two million licensed drivers in the city) and clogged subways had become a regular occurrence at the office of Mayor John ("Red Mike") Hylan. A strong advocate of publicly run subway systems, he denied support and building permits to private operators. (The subway feud would eventually result in Hylan's defeat by the flamboyant Jimmy Walker in the 1926 election.) Bad timing for an uptown eclipse.

Three major snowstorms had hit the city in the first three weeks of 1925, and two more would follow before the frigid month of January ended, the most ever recorded up to that time. (The record stood until 2011.) But this wouldn't deter New Yorkers from plans to view nature's long anticipated spectacle—America's first *urban* eclipse.

On January 24, 1925, the moon's shadow was predicted to hit the ground at precisely 9:01 a.m. on the western shore of Lake Superior, then fly over northern Wisconsin and Michigan, crossing the U.S. border at Buffalo, New York. Speeding across Ithaca and Binghamton, it would arrive, via Paterson, Passaic, and Rutherford, New Jersey, in upper Manhattan at 9:15. From there it would pass over Long Island, New Haven, Connecticut, and Providence, Rhode Island, exiting the continent at 9:17 and heading out to sea for a speeded-up daylong voyage toward takeoff back into space near the Shetlands off the north coast of Scotland (at 10:45 a.m. New York time).

Upper Manhattan lay on the southern edge of the hundred-mile-wide band of totality, which ran west to east, so it lined up almost perfectly parallel to the direction of the numbered streets that ran in the same direction. The most reliable predictions placed the southern line of totality at Central Park and 83rd Street, give or take eight city blocks. That means you'd stand a good chance of glimpsing the corona above 100th Street, the farther north the better. If you could get up beyond Peekskill to Newburgh—a three-hour drive up U.S. 9 in those days, you could net one minute and fifty-five seconds of total darkness. Those positioned below 60th Street would experience only a glimmer of light from a thin crescent sun; in Brooklyn, only twilight.

As the countdown to eclipse day ensued, cities and towns along the band of totality took necessary precautions. Trolley motormen in New London, Connecticut, where more than a thousand students from colleges from various parts of the country intended to locate, were ordered to stop their cars and refrain from turning on their lights during eclipse darkness. Urban visitors were also warned to beware of pickpockets. To guard against possible attempts by highwaymen to commit depredations during the eclipse, the city of Brockton, Massachusetts, doubled the police guard on banks and kept armored cars garaged until mid-morning. The 1920s, after all, were times of rampant criminality. Chicago gangster Bugsy Moran gunned down Little Johnny Torrio on the very day of the eclipse (he survived, but totality was clouded out). Anticipating a sudden electrical overload, the New York Edison Power Company ordered extra tons of coal for the furnaces. What if a million people suddenly decided to turn on their lights during totality? Edison himself joined a party of observers in Westerly, Rhode Island, forty-seven years after first spinning the chicken coop observatory yarn during the Rocky Mountain eclipse of 1878.

On the day of the eclipse, intense cold was ushered in. Dawn temperatures hovered near zero, but the sky was a clear, deep blue. Northbound subway and elevated cars started jamming up early, their southbound counterparts practically empty. People began to gather at street corners that faced east, with cameras, binoculars, and telescopes, with and without tripods, in hand. Those lucky enough to occupy an east-facing apartment window or office evaded numb toes and frostbite. Unable to take the detour north, Midtown people on their way to work stopped at various shop entrances to watch. Children headed to extracurricular activities glimpsed the partial phases with protective equipment supplied by their guardians. Twenty inmates on their way from prison to arraignment were allowed to join their escorts in a brief respite of celestial gawking from the courthouse steps. The observation platform atop the Woolworth Building opened early; it was quickly packed. Observers there reported seeing clusters of people looking like so many little black dots on the rooftops below.

Willy Luyten was one of astronomy's most colorful characters. In his autobiography he calls himself a curmudgeon, because of his habit, like C.H.F. in the Vulcan affair, of finding fault with the research of many of his colleagues.[2] Seeing Halley's Comet in 1910 first turned the eleven-year-old Luyten's eyes to the sky. World War I blackouts in the Netherlands aided his independent search for variable stars, first with binoculars and later with a portable telescope he'd saved up for. On his sojourns to and from his home, with the observing equipment stuffed in his pack, he was arrested a few times on suspicion of transporting stolen goods.

After acquiring his Ph.D. at the age of twenty-one, young Willy immigrated to the United States and took a job at the Harvard College Observatory. There he managed to offend celebrated Princeton astronomer Henry Norris Russell, who was visiting Harvard, by claiming that he had actually made a discovery attributed to Russell. That promptly got him fired. Blackballed from the elite astronomical community at the age of twenty-five, Luyten suffered a further misfortune, an especially damaging one for an astronomer. While playing tennis with the grandson of poet Henry Wadsworth Longfellow, he was hit on the head with a tennis ball and lost sight in his right eye.

Luyten's ability to write accessibly about the workings of science turned the jobless, one-eyed stargazer into a budding science reporter. After a couple of his op-ed pieces on astronomy in the *New York Times* drew praise, the newspaper hired him. With renewed interest in celebrating American scientific achievements, newspapers had given wide coverage to the 1878 eclipse and the confirmation of Einstein's theory during the 1919 event.[3] Luyten was assigned to handle the official scientific reportage on one of the most dramatic public events ever to be witnessed in America's most populated environment—and he would tell his story from an "aeroplane."

Observing eclipses in the air began in 1887, when Russian chemist Dmitri Mendeleev, famous for creating the periodic table of the chemical elements, witnessed totality from a balloon. French astronomers boarded the first eclipse dirigible flight in 1912. David Todd had planned to be first with an airplane to "chase the sun," the first recorded use of the term.[4] He planned to fly at 120 miles per hour for the August 21, 1914, eclipse over Russia. The Amherst astronomer's interest in aerial photography had begun in 1880, when he worked with Samuel Pierpont Langley in experimental airplane design. Realizing the distinct advantage of deploying recording equipment above the surface of the earth, he had already begun to experiment with balloon flights.[5] Sadly, the Great War forestalled Todd's plan. In 1918 the Florida station Todd chose was one of only two entirely obscured during the transcontinental eclipse of June 8. Todd tried again in 1919, this time losing his plane to a hurricane after being grounded in Brazil.

The unfortunate Todd became more eccentric as he aged, and one biographer contends that it was these disappointments that unsettled his mind in later years. He spent the last two decades of his life in a series of nursing homes and hospitals, where he worked on a plan for making life eternal; he called it "vital engineering." He also developed a bizarre theory suggesting that the sun would soon blow up. Convinced of the existence of canals on Mars, during an opposition, or closest approach of the planet to Earth, Todd led an expedition to Chile to photograph them. He also developed a plan, never carried out, to launch a balloon equipped with a receiver at the time of another close opposition to record possible radio signals from Martian life. Later, he persuaded the United States army and navy to listen for the same signals at their radio receiving stations. Undaunted, Todd launched his daring plan to fly above 13,000 feet (that is, above one-third of the earth's atmosphere) to witness the 1925 eclipse. As the *New York Times* explained:

> To neutralize the jar and vibration, the camera is secured in the cockpit by heavy cotton-covered rubber rope. The pilot goes to his maximum ceiling, then cuts off his engine and glides as slowly as possible in a straight line away from the sun. The observer adjusts the finder so that the sun's image will be centered, and exposes several negatives. If carefully secured and focused, the solar image on the film should appear in a succession of small, black and practically circular disks. The film is developed slowly with a weak solution. When the thin streamers or brushes of light make their appearance around the edge of the black solar disk the developing is stopped, as this is a phenomenon which it is desired to show.[6]

Sadly, Todd, by then advanced in age and under orders from his doctor, would miss his final opportunity.

By the mid-1920s, many professional eclipse chasers were eager to leave the rails and soar in the sky, just like the celebrity risk-takers of the golden age of flight—anything to get a good shot of the sun's corona. There had been other failures besides Todd's. Sixteen naval aircraft attempted aerial photography of the September 10, 1923, eclipse, but not a single in-focus image resulted. Today, the record for time spent in the shadow is held by a Concorde supersonic jet from 1973. The jet remained in the shadow for seventy-four minutes, more than ten times the maximum duration obtainable at ground level.

On the morning of January 24, 1925, skies in New York and the vicinity were ready to host twenty-five aircraft, including the U.S.S.

Los Angeles, the world's largest dirigible. Experts on the ground would analyze the sun's spectrum during totality in hopes of reading chemical signatures in the bright lines emitted by the sun's chromosphere; they would also attempt to continue their measure of the temperature of the sun's corona. A pair of twin cameras was set up to re-test the Einstein effect, first revealed in 1919's celebrated eclipse that changed the world.

Science triumphant! Knowing in advance that something will happen and accepting the reasons why has a way of substituting curiosity and keen interest for dread that—as a New York Times op-ed put it—differs "so wonderfully from the mental state of those human beings whom such a natural phenomena used to fill with superstitious fears."[7] More than a hundred pieces on the eclipse appeared in the *Times* in the month leading up to the event, the first of its kind since the city's birth to take dead aim on New York. During the last, in 1478, the great-great-grandfather of Peter Minuit (the man who bought the island from the natives for 60 guilders, today about $1,000) walked the roads of German Westphalia. Manhattan was then a rocky mount thick with forest.

In a dozen articles written prior to the eclipse, Willy Luyten prepped an anxious waiting public on what to see and where and when to see it. He also clearly explained what scientists hoped to gain by viewing the event from airplanes and dirigibles. When it was all over, he would report on what he'd experienced flying with the largest army fleet (thirty-five planes) since the Great War, all directed toward a mission of scientific data collection. Luyten wrote of his experience with a passion not often expressed by a research scientist: "Never before have I been so conscious that I was in infinite space, free from all connections with the earth, watching an event take place between three celestial bodies and feeling myself, as it were, on the same footing with them. All four of us were sailing through space with nothing to hinder us, nothing to stop our path, completely free and separate entities in the universe."[8]

"There it is!" onlookers cried at 8:02 a.m. Then came the long, frigid wait between the hours of eight and nine, as the lunar bite grew bigger and the sun reenacted its eternal imitation of a waning crescent moon. "Save Your Eyes for Ten Cents," read a sign held by a Broadway street peddler, one of hundreds dispensing fragments of smoked glass, strips of exposed film, and pin-pricked index cards for self-styled shoebox pinhole cameras.

Early twentieth-century urbanites appropriately armed with smoked glass/exposed film for observing a total eclipse of the sun directly. (Wikimedia Commons)

Sungazers who forgot to bring smoked glass took to innovating. Bowery workmen smashed soft-drink bottles and hastily smoked the fragments over a barrel fire to peep through. One inventive truck driver lit a match and held it to the windshield of his cab, blackening enough of it to offer a comfortable view. And a clever nurse came to the aid of patients in a Long Island hospital unable to turn to look out the window. She smoked a piece of glass and held a mirror behind it so patients could see the sun's crescent image from their beds.

At the two-thirds mark between first and second contact, a few streetlights turned themselves on and shadows began to sharpen. (As it turned out, power consumption increased only by 20 percent during the daytime darkness.) Back in the city, on the crowded steps of City Hall, Mayor Hylan, witnessing city lights suddenly turn on, remarked: "The sun may be eclipsed, but New York—never!"

One observer reported poetically that a thin sunshine "washed everything on which it fell with forlorn and melancholy tints."[9] It was a twilight "like that by which Charon navigates his ferry boat," suggesting the terror that livid flames "cast pale and dreadful" in the underworld. Three women, terrified by the disastrous twilight in Morningside Heights, fainted. One reporter noted that just past nine, when the light grew very feeble, a few gawkers joked about the end of the world and the anticipated appearance of Jesus. Then the sun suddenly dissolved and the "diamond ring" appeared. "Awed by Jewel of Light Hanging from Luminous Ring," reads a January 25, 1925, *New York Times* headline: "A thin, luminous ring, set with a great gem of soft-burning light, hung in the eastern sky yesterday morning. . . . For several seconds the jewel sparkled with a pure and mild radiance, then trembled and melted into the circle of the light which rimmed the inky disk of the moon. . . . The ring, with its gorgeous solitaire."[10] The analogy stuck.

Astronomers planted volunteer observers along Riverside Drive north of 72nd Street to indicate by street corner whether they actually saw the sun totally eclipsed and, if so, when. Why was this important? Nailing the precise spot to within a few blocks would amount to a difference of less than a mile in updated calculations of the diameter of the moon. One observer at 240 Riverside Drive observed a split second of totality, while another at 130 Riverside Drive saw only a tiny sliver of crescent. They were 140 meters (225 feet) apart—three times the width of 86th Street.[11]

Those fortunate enough to get to the north end of the island variously described the color of the corona as pale lemon, pink, and yellow-green. Careful watchers in the Upper Bronx copped enough time to glimpse the alternating, quivering, dark and light shadow bands on freshly fallen snow. Then the planets popped out, first Mars and bright Jupiter; then, closer to the sun, Mercury appeared. A few thought they saw the bright star Altair.

"Corona Divinely Beautiful: Dr. Luyten, the Times Observer, Describes the Scene from Airplane," read the headline of the Sunday *Times* the day following the eclipse. Why fly to an eclipse? As Willy Luyten explained, the earth's atmosphere is a major hindrance to anyone wishing to collect data. It sucks up radiation from a wide range of the spectrum, and its dusty ambience scatters and diminishes incoming light.

Luyten also reported on the results acquired from enclaves of ground-level astronomers perched at their telescopes all across southern New England. They had worked hard, amassing all the information they could muster in the one to two minutes of totality nature had doled out to them. A major result was that second contact (the beginning of totality) came four seconds late—that made the *New York Times* headline next day. The difference works out to one mile in the position of the shadow. That knowledge, acquired from telegraph signals at stations along the observing path, would provide the wherewithal to recalculate a more precise lunar orbit, as well as a more accurate estimate of the diameter of the moon. Astronomers aloft in their twenty-five aircraft took some of the best corona photographs and spectra to date.

Luyten was surprised to see that the corona extended as far as three times the size of the sun's disk—exceedingly large for solar minimum. Of the disputed mystery element, "coronium," alleged to have produced the green line in its spectrum, the astronomer was ahead of his time, positing that it had to be caused by a known element whose spectrum he hadn't fully realized. Observations during the 1932 eclipse would prove him correct. Regarding the cause of the shadow bands, Luyten explained that since no one in the air saw them, they must have resulted from a low-level effect in the earth's atmosphere. Again Luyten was correct.

Radio transmission anomalies proved more difficult to account for. According to Luyten, the still-correct explanation is that the sun's radiation ionized the air so that it could not transmit long wavelength radio waves. That's why radio transmission is much more effective after dark. Try tuning in distant radio stations at night: I still think fondly of listening to Dallas and Los Angeles radio stations on my battery-operated portable after midnight on the east coast. So, in eclipse conditions AM radio reception should improve. It didn't happen. Instead the signal intermittently waxed and waned. At shorter wavelengths, Luyten explained that the moon's shadow, as a column of darkness, acts on radio waves the way a column of glass affects light waves, producing a diffraction pattern of alternating dark and light zones. That could account for the audibility increase and decrease as the dark shaft moved along.

Why all the interest in radio reception back in the 1920s? Radio communication had begun to play a vital role for combatants in World War I, though all amateur and commercial radio was shut down by the government once America entered the war in 1917. The U.S. Signal Corps developed the first radio telephones, which they introduced in the European theater a year later. With the advent of World War II, radio messaging would prove to be of paramount importance.

In one of his *Times* pieces, Luyten told the exciting story of observers shooting pictures of the corona from a narrow runway atop a giant navy dirigible. Flying at three thousand feet, venturesome astronomers braced themselves against the wind on the tiny track along the top of the craft shooting motion picture film in the five-degree weather. Imagine, says Luyten: less than three hours after landing, footage of the eclipse was being shown in Broadway theaters. (The cost of sending the newsreel films to the west coast set an airmail postal record: $536.14.)

There was excitement at ground level too, where astronomers focused on the flash spectrum of the chromosphere, the layer between the photosphere and the corona, visible at the commencement of totality. Looking through a small spectroscope attached to a pair of binoculars, when the operator saw the flash, he pulled a string to expose 5 × 7–inch sensitized photographic plates in a big spectrograph (basically a spectroscope and camera attached to a telescope); there were nine such apparati.

Luyten would later credit the *New York Times* with helping him secure a job for life at the University of Minnesota, where he spent the next seven decades calculating stellar motions and criticizing other people's work, justifiably about two-thirds of the time. His specialty was white dwarfs, hot faint stars in the late phases of stellar evolution, an endeavor one wouldn't think might possibly get him in trouble, but this *was* Willy Luyten. He had applied for a government research grant for expenses to attend a conference in Scotland on white dwarfs. The surgeon general, in charge of the funding organization, demanded that Luyten provide yes-or-no answers to two questions:

1. Are human subjects going to be used for experimentation in your conference on White Dwarfs?
2. [Do] you realize that Federal funds may not be used for segregated conferences?[12]

With the onset of third contact, the darkest moments of the 1925 eclipse were over. As quickly as it had materialized, the corona vanished, the diamond reappeared on the opposite side of the ring, shadow bands rippled for a minute, and the crescent sun resumed its shape. The partial eclipse show went on, if anticlimactically, for an hour more. Though the ribbon of crowds, which had numbered in the tens of thousands, began to disperse, many stayed on entranced, hypnotized by the spectacle despite numb feet.

There was plenty happening outside the Big Apple during totality. One group of scientists had booked themselves into a Niagara Falls hotel to study the effects of the eclipse on the scenic values of ice and water, part of a plan to develop a colored artificial lighting system for the Falls, which was later implemented.[13] Biologists also investigated the moon's shadow. In the thirty-second period of darkness at the Bronx Zoo, they monitored a herd of deer racing around their range, as birds rushed up tree branches and tucked their heads under their wings, owls flew about, and monkeys quit their usual early morning jabbering. Were they wondering why the day was so short? One of the big alligators used to making long, guttural, vibrating sounds to greet the twilight repeated his performance on what some New Yorkers called the day with two mornings.[14]

Accompanied by their two dogs, President and Mrs. Coolidge took ten minutes to view a partial eclipse from the grounds of the White House. Noting the curiosity of his Airedale with what was happening, the president lifted the dog on its hind legs and held a window-size pane of smoked glass in front of its eyes. The dog lost interest.

Later that day it was reported that one of the Senate barbers went mad. He brandished a razor and began making slashing motions at his customers, shouting that the world was about to end. The

Eclipse viewers of the 1925 eclipse included President Coolidge, his wife, and his uninterested dog. (Courtesy of the Library of Congress)

New York Times also reported an apparent eclipse miracle. In his Hackensack home across the Hudson, a nearly blind man remained content to listen to his children narrate the phenomenon. When they exclaimed "The sun is eclipsed!" he instinctively made his way to the window and looked up. Suddenly racked with "pain" from the glare, he stumbled back to his chair. Two hours later his full vision allegedly returned. "I can see you all as clearly as I did years ago," he uttered. "God be praised!"[15]

In 2009, on the eighty-fourth anniversary of the event, broadcast meteorologist–amateur astronomer Joe Rao reported a conversation he had with a few of the remaining eyewitnesses to the eclipse of 1925: "I don't remember much about the eclipse, so much as I recall that Papa woke me and my two brothers up very early that morning, bundled us up, piled us into our car and took us on what was then (for us) a great adventure. We were going to the Bronx."[16] Another surviving witness sent Rao an eclipse viewer, a piece of exposed film mounted on a piece of cardboard, dated January 24, 1925, with instructions on how to properly use it: "I've held on to this for many years . . . maybe you can put it to some use."[17]

Others remember their parents taking them on an adventurous car trip to the upper Bronx, or standing on a corner, their feet frozen in their boots. But Midtown kids of that age who will witness the big events upcoming (they'll need to head a good day's drive south along I-95 to view the 2017 eclipse, or up I-87 a few hours in 2024) will have an excellent chance of being around on May Day in 2079 for New York City's next total eclipse. In a near repeat of both the 1478 and 1925 events, the shadow again will touch down west of the city at sunrise, this time meting out a solid two minutes' worth of midtown darkness at dawn: no need to head north on the subway this time.

Sorting through the multitude of *Times* clippings on America's most watched eclipse, I sense a feeling of communal unity in the presence of the glory of nature, a sudden pause in the rapid pulse of urban life. All the feelings ever felt during totality are there, anticipation, excitement, fear, awe, though in different proportions from what they appeared to be a century earlier.

CONTACT THREE
Lessons from Eclipses

• • •

(*Overleaf*): Contact III / End Total (© 2008 Fred Espenak, MrEclipse.com)

15

The Eclipse as Cartographer and Timekeeper

At 24 degrees after sunrise, solar eclipse. When it began on the south-west side, in 18 degrees of day in the morning it became entirely total. Venus, Mercury and the "Normal Stars" were visible; Jupiter and Mars, which were in their period of disappearance, became visible in its eclipse. . . . It threw off [the shadow] from south-west to north-east. [Time interval of] 35 degrees for onset, maximal phase and clearing.

—Babylonian astronomer, second century BCE

So, what good are eclipses anyway? It's a question I'm often asked, one I can anticipate in an age when we are prone to measure the value of scientific knowledge by its utility. Today we think of eclipses as marvels of nature just to look at, not to acquire practical information from. As we've seen, this wasn't always the case, and it still isn't.

Look at any world map from the sixteenth and early seventeenth centuries and you'll notice how distorted it appears, particularly in the east-west direction: Florida and Baja California are bent out of shape, and Cape Cod seems too long. Portions of the east coast of North America swerve exaggeratedly east and west. On the other hand, Boston and Miami seem about the right north-south distance apart, as do Chicago and New Orleans. That's because latitudes of places have long been relatively easy to measure. All you need to do is measure the angle of the Pole Star above the horizon (altitude).[1] Or you can calculate latitude by marking the altitude of the sun at noon, then making use of tables that give the north-south distance of the sun from the celestial equator. Using both these methods, my students have been able to determine latitudes to an accuracy within a mile.[2]

Measuring longitude is far more challenging because it's time dependent. The earth rotates 360 degrees on its axis from west to east in twenty-four hours, so one hour of time equals fifteen degrees in longitude. The difference in longitude between two locations is the same as the difference in local time, or the angle of the sun east or west of the meridian (the local north-south line that passes overhead). To determine the difference in longitude between two places, you need a pair of accurate clocks and a precisely predictable celestial event that's visible from both places, like a total eclipse of the sun. That situation didn't happen until the middle of the nineteenth century.

The total eclipse of July 7, 1842, was ideal because its path was due to cross southern Europe; the precise timings of the four contacts from the south of France through northern Italy and Germany could be used to link maps covering cities hundreds of miles apart. Astronomers had already calculated exact times of the anticipated occurrence of each contact at specific latitudes and longitudes along the course of the eclipse path.[3]

The problem of determining longitude at sea, a necessity in securing a map of the world, proved far more challenging because clocks on ships were less reliable than those on land. (Imagine timing the four eclipse contacts with a pendulum clock on a vessel rocking in the waves.) The problem was essentially solved in 1773 by John Harrison, who constructed the first precise marine chronometer, a mechanical wind-up clock with a delicate balance spring allowed to swing freely in its wooden case. Today satellites and atomic clocks have supplanted eclipses as sources of precise data in making maps of the world, but marking the contact times became important for accurately mapping another world, specifically for measuring the moon's diameter and the shape of its orbit.

Precise records of ancient solar eclipses, like the April 15, 136 BCE, event described in the epigraph, offer valuable information on how the world turns. Ever since the invention of the sundial we've marked the hours of the day by following the movement of the sun across the sky, but those hours are slowly lengthening. Records of ancient eclipse observations prove the earth's rotation period is increasing, largely due to tidal friction acting like a giant brake against the earth's surface.

Suppose we could compare a hypothetical clock, one that ran continuously—say, since the birth of Jesus—with a contemporary timepiece. The two should be out of whack, but by how much? The answer comes from watching eclipses, one of the few sky events you can time precisely. The ancient Babylonians did their best with water clocks to aid their penchant for closely tracking beginnings and endings of eclipses. The cuneiform text that goes with this chapter's epigraph gives an astronomer's account of a total eclipse of the sun that took place in Babylon on April 15, 136 BCE.

Since one degree is the time it takes the sun to move through 1/360, or four minutes, of its 24-hour cycle, the eclipse must have started 96 minutes ("24 degrees") after sunrise. Totality began "18 degrees," or 72 minutes, later. The entire eclipse, from first to last contact, lasted for "35 degrees" (140 minutes). Using the rate of our contemporary clock, astronomers can back-calculate and arrive at the circumstances of the ancient eclipse, the exact timings and locales, assuming the length of the day hasn't changed. It turns out that these backward projections always yield total eclipse paths situated consistently to the west (and therefore occurring earlier in time) compared with localities where ancient observers say they witnessed a fully eclipsed sun. The retrodicted track of the 136 BCE event shows it passing through Alexandria, missing the eastern Mediterranean coast and just clipping the southwest shores of Asia Minor, before it veers just off the east coast of the Black Sea. But Babylon, where there are records of people having actually viewed totality, lies well to the east of the calculated path.

Comparing this result with the historical record, astronomers determined an average difference in the three timings of three hours and twenty-six minutes. In other words, the earth's rotation has slowed down by nearly three and a half hours over the course of some twenty centuries. While this doesn't really amount to much per day, if you consider the number of days in two millennia (about 720,000), the Babylonian day must have been about seventeen milliseconds (thousandths of a second) shorter.[4] Taken over the long run, this can add up. At the accepted slowdown rate, we can expect to have twenty-five-hour days by the year 152,650,000. Can we sleep in that extra hour?

More recent solar eclipse observations have turned up a few eye-catching surprises about the object being eclipsed. A major one is that the sun may be shrinking. In 1979, two astronomers measured the time it took the sun's image to cross a fine vertical wire fixed in the eyepiece of a telescope.[5] Comparing that with observations recorded as far back as 1750, they concluded that the sun's diameter had shrunk by two arc seconds, about 0.1 percent per century. Percentage-wise that isn't much, but it translates to an alarming 900 miles in 200 years, a result that caused a sensation in creationist circles.

Number 85 on the list of 116 "categories" of evidence for mechanical engineer Walter Brown's "The Scientific Case for Creation" is the

shrinking sun phenomenon.[6] Extrapolating the astronomers' findings, he concluded that the sun would have been large enough for its
surface to have reached all the way to the orbit of the earth a mere
20 million years ago. So if the sun had existed even a million years
ago, it would have been so large that life on earth could not have survived because of the extraordinary heat it would have emitted.

Those dedicated to the fundamental tenets of contemporary biblical creationism believe that the earth is *not* billions, nor even millions, of years old. They follow conclusions reached in the seventeenth
century by James Ussher, archbishop of Armagh, Church of Ireland,
who calculated that creation began on October 22, 4004 BCE, at
nightfall. His method consisted of using the lengths of kings' reigns
recorded in the Bible; the result differs only by a few years from similar computations made by earlier scholars who followed the literalist
tradition, among them Kepler and Newton.

On the contrary, in the late nineteenth century, geologists concluded that it must have taken millions of years to create geologic
formations like the Grand Canyon. Meanwhile, astronomers studying the origins of stars believed gravitational collapse (the infall and
consequent heating of gases from an extended parent cloud) was the
only viable source of solar energy. It was well known, even in the late
nineteenth century, that such a store of energy could only keep the
sun shining for about 20 million years. This conflicted with geological
studies that dated terrestrial rocks to ages much older than the sun.
The discovery in the 1930s of thermonuclear fusion as the source pow-
ering the sun and stars for billions of years resolved this chronological
dilemma, at least for the scientific community, by yielding an age
of the sun equal to or greater than the accepted age of the earth of
4.6 billion years. Ever since Darwin, creationists have continued to
view the scientific community's belief in the vast age of the universe as

having been driven by a blind faith in the theory of evolution. Of the shrinking sun theory, physicist and creationist Hilton Hinderliter wrote, "It is clear that we have witnessed a major scientific defeat for evolutionism."[7]

The shrinking sun episode is a good example of what happens when you uncritically accept a short-term observation and extrapolate it linearly backward in time. Worse still is that the creationists have ignored follow-up studies. Scientific critics of the astronomers who thought they had detected a shrinking sun pointed out that there's a lot of error bias in such observations; the brightness of the sun and the steadiness of the earth's atmosphere complicate things further. You can't reliably extend a single measurement backward to conclude very much about how big the sun was in the past.

Here's where solar eclipses enter: compare contemporary timings of the duration of totality with those dug out of historical records. The earliest eclipse with times precise enough to reach a conclusion took place in England on May 3, 1715, when eclipse watching was big business. Recall that astronomer Edmond Halley had distributed maps describing eclipse tracks and boundaries over southern England. He netted eight acceptable responses yielding a duration of totality of three minutes eighteen seconds, to the nearest second. If we convert these timings to a measurement of the sun's diameter, we get a result 0.2 seconds of arc *smaller* than the contemporary accepted value. Does this mean the sun is actually *expanding?* Unfortunately, the errors of measurement turn out to average 0.4 seconds, which is twice the size of the result. (That would be like concluding that, based on several measurements of my height, I am six, give or take twelve, feet tall.)

Oddly enough, the most accurate data about the sun's diameter comes from observations made at the edge of the path of totality, where the moon's disk just grazes the sun. In effect, because the distance and diameter of the moon are known very precisely, astronomers use measurements of the size of the moon's shadow to compute the size of the sun. Here the timings depend on a knowledge of the lunar profile, the jagged edge that passes across the sun as seen from different locations on earth over a series of eclipses. Analyses from more recent eclipses lead to a result of 0.008 percentage decrease, with an error about the same magnitude. But the fluctuations may follow a pattern. At this stage we can't be sure of what's up with the sun's waistline, but the conclusion seems to be that, at least over the past three hundred years, it is both shrinking *and* expanding: the sun's diameter oscillates over a period of about eighty years by 0.025 percent, a little over two hundred miles. As is often the case in science, sometimes you get a result you wouldn't have expected. Timing eclipses to study solar oscillations still seems worthwhile to me, but not in relation to the young earth theory.

In 1979, the last total solar eclipse visible on any part of the United States mainland touched down near Portland, Oregon, a mere fifty miles south of the entry point of the 2017 eclipse track; then it veered sharply north through largely unpopulated parts of northern Idaho and Montana before crossing into the Canadian provinces. (Portland got clouded out.) Twenty Oregon State University chasers drove 150 miles to Madras for a prolonged view of second and third contacts to determine the precise shadow boundary. Maps of totality zones were still in error by several miles. By stationing observers at 400-meter (437-yard) spacings along a north-south road (just like the volunteers at city block intervals in Manhattan in 1925), they found the graze line on the southern side of the totality band to lie four miles north of published predictions.[8] They dutifully submitted their result to the U.S. Naval Observatory Collection Center.

Broadcast journalist Walter Cronkite devoted five minutes of his prime-time nightly newscast to the last total eclipse of the millennium to touch the mainland.[9] Part of the segment was devoted to neopagans visiting a Stonehenge replica in Goldendale, Washington, who invoked their gods to keep the skies clear (the clouds parted briefly for a momentary view). Fireworks and roman candles were launched. "I was scared to death," responded a young interviewee. "It's a religious experience," remarked another.[10] ABC countered with a news special that included live coverage.[11]

Today there may be less interest in mounting ground-based scientific eclipse expeditions to collect eclipse data since the days of Einstein's eclipse, but astronomer Jay Pasachoff is among the most ardent proponents of the need to continue to do so. He is a strong advocate of ground-level observations of the corona.[12] Observing from the earth's surface is far less expensive, and more flexible, he argues, and it yields higher resolution data than what we get from aircraft or satellites.

That the corona has a substantially higher temperature (two million degrees) compared with that of the sun's surface (approximately six thousand degrees) is one of the major findings to emerge from solar eclipse studies. Once spectroscopic studies, begun in the 1830s, led to the identification of highly ionized iron and calcium in the bright-line spectrum a century later, it became clear that the corona was extraordinarily hot. It doesn't take a lot of radiation to power up the corona because it has a very low density. One theory has it that the sun's outer atmosphere heats up because the low-density gas, or plasma, that comprises it is resistive to the dissipation of energy.[13] Precisely what causes all the excess energy to get up so high into the sun's atmosphere is another unsolved problem. The direct connection between sunspots and coronal streamers was known as far back as the

1890s. It wasn't until the mid-1930s, when the extent of the solar magnetic field was mapped, that astronomers began to theorize that ionized particles moving along magnetic force lines were at least partially responsible for the energy transfer. By monitoring the corona at different wavelengths for intensity variations during a total eclipse, ground-based observations reveal a strong relation with upward energy propagation and the sun's magnetic field, though solar physicists still debate the physical mode of transport.[14]

Do other stars similar to the sun also possess coronae? Data from X-ray telescopes suggest coronae are common features, many of them more extensive than the sun's, in the outer atmospheres of most stars. The behavior of the hot, low-density gas that makes up the sun's corona offers us a close-up laboratory for studying the production of X-rays and the interaction of hot plasmas with magnetic fields in stars similar to the sun.

A final reason to study eclipses is that we actually live in the outermost reaches of the sun's atmosphere. What happens in the corona may explain changes in our own terrestrial environment. The solar wind, or outward streaming of solar particles, comes to earth via expansions of the corona, which can cause power outages and affect earth's climate.

So eclipses are worth something. Of course, for our ancestors there was much more at stake than taking the measure of the lunar and solar participants—and the animate world still responds instinctually to life in the shadow.

16

Zoologists Chasing Shadows

Where nature is largely free from mental trauma, the opposite is
the case with humankind. Not only does the experience of an
eclipse make an indelible experience upon all who share it, but humans, with their remarkable and unique gift of speech and language, can communicate it to others of their kind.

—*Thomas Crump, 1999*

I am not sure I agree with the first part of Thomas Crump's statement. Eclipse stories include anecdotes about how animals respond to
nighttime's unanticipated onset. Birds fell to the ground, according to
a 1560 eclipse account. In 1842 a starving dog refused to consume a
crust of bread until the end of totality. And during the July 28, 1851,
eclipse in Sweden, ants busily carrying material to the nest stopped
and remained motionless during the darkness. Verifying such inci-
dents is difficult because they are usually reported casually, most often
by astronomers busily tending to what's going on with the sun.

Even if they paid attention, astronomers know no more about
observing animal behavior than do zoologists about what to watch for
in the sky during a total eclipse of the sun. Untrained observers usually assume that an animal would be frightened or disturbed by an
eclipse and that its unorthodox behavior could be interpreted as a response to a stimulus. That is easy to conclude if you're not familiar

with say, the habits of birds. "Just as the darkness was most intense, a male crown crane spread its wings and ran across the cage with head held low and wings flapping," remarked one zoo spectator during totality. The zookeeper responded: "Not an hour goes by without such a demonstration on the part of one or several cranes in the cage."[1] So let's look at solar eclipses through the eyes of those who study the animal kingdom.

It seems that when an eclipse happens, animals active in the daytime may try to do what they normally do when it begins to get dark. Squirrels start scurrying up and down tree branches seeking shelter, dragonflies hide beneath leaves, sheep and cows quit grazing and head for the barn, and birds quit chirping. During the 1851 eclipse, a singing canary stopped at the onset of totality and remained stationary on its perch. Meanwhile, hundreds of swallows and sparrows outside "flew about like mad things, seeking trees and bushes as places of concealment as if to remain in the air."[2] "Pigeons being fed were much alarmed and disturbed, stretching their necks upward to the sky as if apprehensive of some bird of prey," wrote a Portuguese observer during the May 28, 1900, eclipse.[3] As totality approaches, blackbirds and roosters, who normally assist us in and out of slumber, commence their customary chatter to greet twilight. During the total phase, nocturnal creatures, like crickets, pick up where the birds leave off; owls hoot, bats fly out of their caves, and moths, who use nocturnal light for navigating, begin to flutter.

Even though such anecdotal narratives go back centuries, I was surprised to discover that very few scientific studies of animal behavior during eclipses have actually been conducted. Most are buried in professional biological journals. One of the most thorough investigations took place during the eclipse of August 31, 1932 (total in eastern New Hampshire and southern Maine), when Boston Society of Natural

History zoologists set up trained observers to monitor animals ranging from insects and cold-blooded vertebrates to birds and mammals at selected locales in farming districts in eastern New England.[4]

It took an individual with considerable cross-disciplinary interests and connections to put together such a massive effort in the 1930s, but Harvard entomologist William Morton Wheeler (1865–1937) proved himself equal to the task. Sent off at an early age to a strict German academy in Milwaukee due to what he described as his "persistently bad behavior," Wheeler quickly fell under the yoke of academic domestication: "To have annoyed one of the burly Ph.D.s who acted as my instructors, as I had annoyed the demure little schoolmarms in the ward school, would probably have meant maiming for life at his hands or flaying alive," he said.[5] A museum attached to the academy sported an impressive collection of stuffed animals, snails, shells, and lots of bugs. Wheeler fell in love with them. He spent late-night sessions tirelessly mounting and classifying insect specimens. With more than four hundred titles to his credit, Wheeler would become a leading authority on the social behavior of insects. He did his work at a time when scientists were just beginning to entertain the notion that lower animals were social beings; thus Wheeler planted the seeds of sociobiology (Alfred Kinsey was one of his students). Though he was never a status seeker, his efforts would earn him a curatorship in zoology at the American Museum of Natural History and election to the National Academy of Sciences.

Being an astronomer, I cannot imagine the dedication and sheer willpower it takes to focus your attention on an animal during a total eclipse rather than on the sun. Notwithstanding, the well-trained cadre of Boston biologists brought together by Wheeler dutifully collected their data. They found that, as you might expect on a late summer night, crickets began to chirp noisily at about fifteen minutes before totality, by which time the intensity of sunlight had fallen by 90 percent. In some locales, the chirping lasted up to an hour after the one minute and forty-five seconds in the shadow. Many observers characterized the noise as extremely loud, but the biologists couldn't be sure how much of the noteworthiness attributed to it came from the observers' expectation and concentration of attention to it.

"Mosquitos emerged from the grass and were annoying during the darker portion of totality," one of the expedition's entomologists noted. "Mosquitos came out and I was bitten," wrote another.[6] Gnats began milling in swarms, and a grasshopper remained motionless at the top of a sunflower. A member of the team reported that he was "informed by a lady that shortly after the total eclipse her pantry was greatly infested by cockroaches."[7]

The Boston study also included fishery supervisors. They recalled that trout ceased feeding as soon as it grew dark. But goldfish that usually dine in the evening swam to the top of their pool. One got so hungry it decided to feed off the tail of a companion, killing it. Always ferocious, pickerel started jumping in the Charles River, and lake bass in New Hampshire bit so avidly during eclipse twilight that one fisherman reported catching twelve of them; however, a minute before totality, they suddenly stopped biting. On the other hand, another angler complained that none had gone for the bait during total eclipse, nor had any bitten that entire day. Among the amphibians, peepers, or tree frogs, sing at dusk, especially on rainy days in New England swamps. They came out on cue, at first intermittently at 3:45, then becoming more vociferous at the 4:30 eclipse maximum. Turtles left their sunning spots promptly at 4:25 and slipped into the lake but returned in unison to their posts by 4:40.

Evidence collected from the New England expedition led to the conclusion that all birds showed no reaction to the eclipse in regions where totality was 98 percent or less. The zoologists characterized reactions during full eclipse as "fear, bewilderment, though not terror or panic"; but how do you express animal feelings in human terms?[8] Birdwatchers saw no evidence that their subjects had sensed the impending darkness in advance, and there was no difference in behavior across avian species, including domestic chickens, ducks, and caged birds. Sea birds may have been an exception. "At the moment of totality," reported one shore observer, "every gull on the island, nearly, went screaming into the air . . . and flew aimlessly about as long as the sun was hidden."[9] On the other hand, another flock only engaged in increased sound activity. So maybe it depends on the bird?

Do chickens really go home to roost, or is the story of Thomas Edison and the 1878 eclipse all made up? The Boston study found that, as with the shore birds, it really *does* depend on the chicken: some do and some do not. Two observers reported that their hens, having acted excitedly minutes before, suddenly bolted for the coop at the moment totality set in; but others noted that theirs stayed put. Is there a chicken psychologist in the house?

Reactions of different species of birds to different levels of sunlight varied greatly. Fifty percent of the way to totality during the eclipse of February 26, 1998, royal terns that had been hunting in a bay on the coast of Venezuela suddenly left the area. At 85 percent totality, frigatebirds ascended nearly three hundred feet and flew inland. At the same time, pelicans flew in single file to their cliff roosts, and gulls formed a loose flock on the surface of the water.[10] During totality not a bird was in sight, except for the gulls who flew tightly packed and erratically. At 70 percent partial following totality, the pelicans and frigatebirds returned and resumed normal activity. The terns did not. Maybe their departure wasn't stimulated by the too-low light level? On the other hand, during the 1997 total eclipse in India, black-crowned night herons increased their calling, preened excessively, and shifted on their perches.[11]

During the 1932 eclipse, dogs in an animal rescue center were quieter than usual, though some were said to have been frightened; for example, a Chow pup ran and hid under a shed, while an Irish terrier whimpered. Another dog (breed unknown) seemed scared and cowered at the side of its mistress. Dogs yelped during totality in several instances. But again, as with the chickens, in an almost equal number of cases there were no noticeable canine reactions. Observations of cats yielded no abnormal behavior either, with one exception. In York Beach, Maine, a Maltese meowed at the onset of twilight, then let out a shriek and rushed up a post as the corona appeared, strangely riveting its eyes on the sun all the while.[12]

Wheeler's Boston biologists collected observations of twenty-one species, about half of which (red fox, mink, raccoon, harbor seal, cottontail rabbit, woodchuck, muskrat, beaver, wapiti, and goat) exhibited no reaction. Species most affected were dog, bat, rhesus monkey, skunk, gray squirrel, deer, domestic sheep, and cattle. The accounts that dealt with the latter two caught my attention: "At Lisbon Falls, Maine (totality area) a flock that was grazing peacefully became disturbed as darkness came on, and at exact totality the entire flock stampeded toward the barn; a flock at Conway, New Hampshire, also came in to their evening pens, bleating, in a bunch as they were accustomed to do for the night, but seemed in no way excited."[13]

The animal scientists debated what had happened and came to the conclusion that the varying behavior may have been motivated by a difference in habit among the various groups. Some were accustomed to return to shelter at evening, while others tended to remain longer in pasture: "Again the response to one or two individuals in a group may have affected the actions of all as in 'stampeding' for the

barn. At North Conway, New Hampshire, cows were mostly uncon-
cerned, but in 29 of the 39 cases, however, there was a quite definite
response to the oncoming of darkness, in most instances a starting in
the direction of the barn, a feeding toward the barn, or arriving at the
bars of the pasture or barnyard ahead of the usual milking time."[14]

What surprised the dedicated life scientists who participated in the
landmark 1932 study most of all was the variation in eclipse responses
within individual species. Animal watchers found behavioral differences *within* a species to be greater than those observed *between* two
different species. What could account for some gulls flapping wildly
while others paid no attention, or some chickens scrambling to get to
the coop while others clucked along apathetically? Like humans, some
birds may be more fretful than others. This was a striking revelation in
1932, a time when animals were less regarded as sentient beings. Today
animal psychology is a burgeoning field in the veterinary profession.

Because there were many apiaries in the region where the 1932
study was conducted, beekeepers, who were well acquainted with the
normal behavior of the bees, registered significant observations:

> I have fifteen colonies of bees which I watched closely. The
> field bees had been working heavily on goldenrod all day and
> everything was normal till darkness began to come on, when
> they came home in unusually large numbers. When it was
> darkest they had not all reached the entrances of their respec-
> tive hives and it was then too dark for them to see their way,
> so they kept flying about in the air or landed in the grass, till
> it lighted up again. Then they found their way home and
> became very quiet. Some stragglers still came in from a dis-
> tance. Later, after the eclipse was nearly over, they ventured
> out again very slowly.[15]

In stark contrast, another bee colony appeared to have been tricked by the sudden diurnal variation:

> I have four hives of bees. The day was a good one for them to be out gathering honey and therefore when the eclipse began bees were going out and returning at a lively rate. With the first perceptible dimming of the light there was no change in the bees' activity, but as darkness increased the outgoing bees diminished in numbers and the return batallions grew larger. When the light was almost at the dimmest point no bees were leaving the hives, but the returning individuals were pouring in by the thousands. The space above the hives was like a funnel into which bees were literally pouring from every direction. When the light began to increase there was not a bee to be seen in the air. For them evidently it was the sudden advent of night. I watched to see whether with the return of light they would go out again. Normally they go out till dusk, but in spite of the bright sunshine following the eclipse only an occasional bee ventured forth.[16]

For the keeper at another apiary the trick event was similar, but it generated even more erratic behavior:

> During the period of the eclipse today I closely watched the behavior of eleven strong colonies of bees. They have been particularly busy on the big late honey flow from yellow goldenrod and buckwheat, and were very active at the beginning of the eclipse at about 3:30 o'clock. The temperature was 85°F, and the sky partly overcast, with the sun shining through now and then. Here at Lincoln, where these observations

were made, the clouds thickened and only a short sight of the sun at 4:30, just at the height of the eclipse, was obtained. There was no direct sunlight afterward. At 4 P.M., not much change in temperature or in the activity of the bees could be noted. At 4:10 P.M. many more were coming in than were going out. At 4:20 P.M. the air was full of returning bees. Those leaving the hives flew about on erratic courses and came back. They also became excited and cross and it became dangerous to stay within forty feet of the hives. I beat off two attacks and returned to a safe distance. At 4:30, the period of greatest darkness, the fronts of the hives were covered with bees all trying to get in at once. At 4:40 a few stragglers came in—those caught in the dark a long way from home. At 4:45 there was not a bee in sight, not a sound in the apiary except the hum in the hives that is usually heard at night when the ear is held close to the hive. There was no outside activity, all having apparently arrived home.[17]

Family quarrels accompanied a fourth report:

Hive No. 1 consists of hybrid bees. The queen was bought for a pure Italian, but apparently she isn't. Black bees are more nervous than the Italians and this particular hive has given me trouble by robbing the others when the clover flow stopped in July. Either the bees were upset and this started them driving out the drones or, by some sense we know nothing about, they decided that the short Fall days were coming and it was time to get rid of the drones.[18]

Based on these detailed descriptions, the zoologists are asking a basic question: are the animals responding to a built-in clock or to external cues? Of all the rhythmic time cycles affecting biological organisms, those of the honeybee are the most well known, thanks to a once controversial work called *The Dancing Bees* published in 1953 by Austrian entomologist Karl von Frisch (1886–1982).[19] Initially trained in biology and medicine, von Frisch served as professor of biology at the University of Vienna, until he was abruptly dismissed in 1933 on the grounds that he was one-eighth Jewish. He meandered about the academic world until he acquired a postwar position at the University of Graz, Austria.

Von Frisch initiated his pioneer studies of the social life of bees quite by accident. One of his colleagues told him that whenever he breakfasted on his terrace, he noticed that the bees seemed always to be there at the right time. Were they anticipating the savory taste of the jams and marmalades the professor had often set out? They showed up whether or not sweets were at the table. Von Frisch wondered, how does a bee acquire such punctuality? Is there within each animal a sort of internal clock that goes off at the right hour, or does the insect receive clues from the external environment?

To find an answer, von Frisch set up a series of experiments. He placed food sources in different directions from the hive at different times of the day. When bees coming straight out of the hive darted immediately to the food source, it became clear that information about the location of that source was somehow being transferred back to the hive by incoming members of the bee community. When von Frisch peered inside the hive, he discovered that the transfer mechanism was a kind of round dance performed by a foraging bee shortly after it arrived back at the hive to dump a load of nectar. The bee would move rapidly in a narrow circle, completing half a course in a clockwise direction. It would then run along the diameter and finish the other half loop in a counterclockwise sense, making a sort of figure eight. As the bee danced, spectators tagged along after it in great excitement, touching the tip of its wiggling abdomen with their feeders as they followed. Cleaning itself off, the forager-communicator then left the hive and returned directly to the food source, joined moments later by attentive comrades.

What information had been conveyed through this curious round dance? von Frisch wondered. How did the bees who watched the dance know where to proceed? These questions continued to puzzle von Frisch and his coworkers for a long time. Surely, it couldn't be the sense of smell that caused the bees to arrive at their target; they showed up regardless of whether the experimenters laid out honey or sugar water. And the bees certainly couldn't *see* the dancer perform since the comb where all the action took place was sealed in darkness. Any information about orientation to the food source must have been picked up by other means—perhaps by touch, inside the hive. The experiments continued.

When von Frisch varied the distance to the feeding place, he discovered that the farther away the source, the longer it took the dancing bee to complete a loop, from forty runs a minute for a source a few hundred yards away, down to just a few turns a minute for a feeding place situated a few miles from the hive. And if the source was nearby, the communicator bee seemed to wag its abdomen much more rapidly during the straight diametrical portion of the dance. Did this action reflect that the dancer had consumed more energy when running a long way from the hive? Maybe a tired bee automatically conveys information about the distance to a food source because it has less energy in reserve and consequently performs a much slower dance. But what about direction? If the companion bees knew how far to fly, how did they know which way to take off?

Ultimately, von Frisch discovered that the bees were navigating by the sun: they had developed a brilliant way of measuring the angle between the food source and the sun as projected from the hive. On a horizontal comb surface, the angle between the straight portion of the dance and the bearing of the sun is the same as the direction between the sun and the source as seen from the hive. In other words, if a food source lies forty degrees to the left of the sun as viewed from the hive, then the bee conducts the straight portion of its dance at a forty-degree angle to the left of the direction of the sun, which, it will be remembered, cannot actually be seen from the dance floor.[20]

If bees know when to turn up at a feeding place, they must possess some sense of time. Furthermore, because they convey information by marking off the angle of the sun, they may be cued in to this timing sense directly by the sun. The bee learns that when the rays of the sun arrive from certain portions in the sky, it's time to forage. Von Frisch even managed to train his bees to become attuned to several different foraging times each day. But this idea of environmental cueing has its problems because the bees, when totally isolated from all outside periodic changes, continued to behave the same way. They appeared at a feeding spot at the time of day for which they had been trained, even when they were exposed to constant artificial light, temperature, and humidity. After years of experimentation with time sensing in the honeybee, von Frisch concluded that "we are dealing here with beings who, seemingly without needing a clock, possess a memory for time, dependent neither on a feeling of hunger nor an appreciation of the sun's position, and which, like our own appreciation of time, seems to defy any further analysis."[21]

How you're trained in science matters when it comes to what you conclude about problem solving. The *Dancing Bees* was greeted with skepticism in the 1950s by insect specialists who argued that bees locate food solely by odors. Von Frisch's detractors specialized largely in mathematical and statistical studies; they examined group behavior and interpreted subjects as simple stimulus-response organisms. On the other hand, von Frisch looked at individual bees. Like William Morton Wheeler, his broader interests centered on the physiology of the senses and its role in the social life of animals and their language of communication. Are bees communicating animals like dolphins, birds, whales? Von Frisch interpreted a bee's round dance as a form of symbolic communication.

The battle of the bee dances was fought on the scholarly turf of the professional journals for more than a decade before von Frisch's explanation of the bee's extraordinary sensitivity to what's going on with the sun won the day. His biographers make clear, however, that the selfless entomologist sought neither war nor victory. Von Frisch was awarded the Nobel Prize for Medicine in 1973.

My honeybee digression should help us understand why the bees in the Boston study reacted so strongly when the lights went out on August 31, 1932. But where is the elusive rhythmic clock that keeps track of bee-time? Nearly all living organisms operate on cycles of about a day's length. Called *circadian rhythms* (from the Latin *circa,* "about," and *dies,* "day"), they have been recognized in Western culture at least since the time of classical Greece. Twenty-four centuries ago, Aristotle observed that certain plants raised and lowered their leaves on a regular day-night schedule. It wasn't until the age of controlled scientific experimentation (twenty-two centuries later) that science began to probe the detailed nature of this biological clockwork. In 1729, the French scientist Jean de Mairan conducted the first controlled light-dark experiment on plants in a laboratory. He found that the daily periodic oscillations of plant leaves persisted even when the plants were isolated from the natural environment. This so-called de Mairan phenomenon has been observed in practically all living forms, from humans to one-celled organisms. In all cases, the cycle is close to, but never precisely the same as, the earth's rotation period (generally it varies between twenty-three and twenty-seven hours for different kinds of organisms). Furthermore, most subjects can be trained after several days to adapt to a new artificial period through environmental control.

Biologically speaking, there is little to be gained by marching to the tune of a different drummer, so it shouldn't surprise us that most biorhythms are virtual duplicates of nature's basic periods, the day, month, and year, all controlled by the sun and the moon. So let me rephrase the biologists' internal versus external time question: How is the connection between life's circadian and celestial rhythms established? One theory, the *endogenous* hypothesis of internal timekeeping, suggests that because these rhythms persist when the organism is deprived of functioning within the natural environment, every piece of living matter must acquire its own timer. (I cheat by having one implanted in my cell phone.) In other words, anything that lives has the capacity to develop its own internal, chemically based timing system. This idea makes sense because the theory of evolution teaches us that the mechanisms of natural selection favor the survival of the organism that achieves an adaptive advantage. Having evolved over millions of years, well-entrained animals know when to keep their mouths open to get fed; the rest lose out. Every successful class of organism needs to inherit and further develop an accurate biological clock so that it can learn when to anticipate environmental change better than its competitors.

In stark contrast, the *exogenous* hypothesis of time sensing, as the name implies, argues that all biological timing depends on outside stimuli. Organisms oscillate with natural geophysical frequencies because they respond directly to changes in forces of an all-pervasive environment. While lab experimenters may think they have totally isolated their subjects by shielding them from light and changes in barometric or thermal conditions, these organisms have subtle ways of sensing what's really happening in the world around them. For example, they might be responding to changes in the geomagnetic field or electrical charges in the air, to shifts in the intensity of background radiation, or even to tides in the earth's atmosphere that rise and fall on a daily or monthly basis. We already know that honeybees have good gravitational sense because they make use of it in their navigational system. How else could the forager who dances on the vertical comb convey the location of food to its hivemates unless it were able to translate the angle between food source and sun into the angle between food source and the vertical? The bee substitutes the direction of the pull of gravity for the direction to the sun. Time, the source of all rhythm, must originate from the outside: time sensing is hidden away in that secret communication between animal and nature that we casually dismiss as "instinct." So say the exogenists.

Today even the most stubborn internalist would concede that there are some exogenous effects on biorhythmic activity; and the tenacious externalists, who in today's world of genetically based research have been relegated to an active, vocal minority, have been forced to admit that some sort of biological clock does indeed exist within us all. Still, the environment is the ultimate source, the so-called *zeitgeber*, or giver of time. It is the environment that entrains all biological organisms to oscillate in the first place. Tide, temperature, season, sunlight and moonlight, all are part of the primal input that drives us to execute our clockwork, both chemical and genetic. Whether time sensing in the animal world is better explained internally or externally, whether it is a direct response to nature or has been nurtured over countless generations of evolution, our peek into the complex world of the honeybee teaches us that all living beings, regardless of where they stand on the ladder of biological complexity, respond to environmental change in surprisingly sophisticated and detailed ways.

So what does the diversity of animal reactions during a total eclipse of the sun contribute to the discussion of whether responses originate from within or without? Is animal behavior during an eclipse triggered by an internal clock or by light cues? Looking back on the Boston study nearly a century later, biologist Stephan Reebs contends that if all the animals in the 1932 study were passive or just got nervous, then the internal clock is more likely behind their behavior. But if they all changed what they were doing—that is, if all chickens came home to roost—light cues would be a more likely explanation. From an analysis of the 1932 studies, Reebs concluded that *both* mechanisms seem to be operating to a certain extent and that they are involved to different degrees in different species.

To shed some light of his own on the problem, Reebs set up an illuminated fish tank in his lab and filled it with cichlids, an African species frequently kept in small aquaria and known for their behavior in retrieving their fry. To keep the offspring safely in the dark, as night approaches parents catch their offspring by mouth two or three at a time and divvy them up among shallow pits they had previously excavated in the aquarium gravel. They continue the process after dark if they haven't completed it. Reebs kept one sample in the lab under a regime of twelve hours lights-on/twelve hours lights-off, dimming the lights fifteen minutes prior to lights-out. Sure enough, the cichlids immediately started the retrieval process when the lights went out. When he tried dimming the lights at six hours (that would be noontime in the tank), or in mid-afternoon, the parents did nothing. Reebs interpreted this as a good example of an internal clock telling the fish it couldn't possibly be twilight. However, when he turned out the lights completely, no matter when, the parents immediately initiated the retrieval process. Like von Frisch with his bees, Reebs concluded: "And with this we have reached the same conclusion as with the observations of other animals during eclipses. It is not black and white."[22]

From highest to lowest order, eclipses affect the animal world. Tiny marine animals, like zooplankton, move from one level of water to another depending on their basic needs, like eating and mating. They, too, respond to the changing light level that accompanies the onset of totality.[23] During the eclipse of June 30, 1973, off the coast of the Cape Verde Islands, marine biologists used echo-sounding apparatus to measure their distribution at various depths of the ocean on the night preceding the eclipse as well as during totality. They found differences. As totality approached, the zooplankton layer began a gradual upward migration, then it moved slowly downward following the return to the partial phase.[24] The analysis is too complex to review here, but marine biologists concluded that an *endogenous* rhythm was responsible, so that while a single disturbance in the *zeitgeber* can't reset a biological clock, the disturbance in the normal light cycle appeared to have disrupted the entrained *exogenous* behavior. Unfortunately, no biologist to date has had the luxury of being able to spend several days watching zooplankton behavior before and after an eclipse to observe further changes.

Animals at the opposite end of the spectrum of life execute this same up-and-down motion in response to the eclipsed sun as the lowly zooplankton. During the annular eclipse of May 30, 1984, chimpanzees at the Yerkes Regional Primate Research Center in Atlanta migrated to the top of a climbing structure as the sky began to darken and the temperature started to drop. As it got darker, they oriented their bodies toward the sun, turning their faces upward at maximum.

When daylight began to return, they climbed down and resumed normal activities. They didn't repeat the behavior at sunset, nor had they done so at nightfall on the previous day, indicating that this unusual event had influenced their behavior.[25] I was struck by the behavior of one young chimp at the moment of greatest darkness: from a treetop, he raised his hand and gestured in the direction of the sun. I have done the same.

CONTACT FOUR
Personal Eclipses

• • •

(Overleaf): Contact IV / End Partial (© 2008 Fred Espenak, MrEclipse.com)

17

Eclipses in Culture

Given the subject matter we study, it seems only appropriate that
while most of us still define ourselves as astronomers and scientists,
we should build our sturdiest bridges across the gulf separating us
from the interested community of cultural anthropologists.

—*Anthony Aveni, 1979*

In contrast to the kinds of questions astronomers ask about their
predecessors, questions that address the things and processes in the
sky, those addressed by the archaeologists and anthropologists focus
on what another culture's astronomy tells us about its people's religion, their beliefs in an afterlife, their politics, and their concept of
history. Thus, cultural astronomy has more to do with what people
believe about celestial happenings than the happenings themselves.
Rather than simply labeling their beliefs religious, astrological, superstitious, or nonscientific, taking *their* perspective seriously allows us to
hold up a mirror to nature, and to see our faces in that mirror set
against a background of other faces, and the minds behind them,
minds that construct worldviews and meanings other than our own
that make sense to them. Drawing constellations or star patterns, following the movement of the planets across a zodiac, or being dazzled

by a solar eclipse, experiencing the sky is not the same for all people. This is one of the wonders of human diversity.

With due apologies for seeming narcissistic, the epigraph above appeared in my summation of one of the first organized conferences on archaeoastronomy in the New World, held at Santa Fe, New Mexico, in 1979. I thought most of the contributors—astronomers, engineers, and other scientifically trained scholars—were asking too many astronomy-type questions and paying too little attention to the cultural aspects of ancient skywatching. Can you blame them? They weren't trained in cultural astronomy; it didn't yet exist. Applying the thoroughness and rigor acquired in their scientific training, they were measuring and analyzing architectural alignments, looking for matches with the standstill positions of the sun and the moon as Hawkins did with Stonehenge. There were too many alignments and not enough people.

The motives of these nascent cultural astronomers also struck me as being colored by a mixture of romanticism and scientific cultural superiority. Their thinking went something like this: Here were people who lived at one with what nature had bequeathed them. Though deprived of technology, they managed to chart out the workings of the heavens with remarkable accuracy. Ah, but what a pity they failed to take the next step and follow the pathway of progress toward rational science. Why did they opt instead for allegiance to an exotic panoply of deities? Little wonder they weren't aware the world is round and that it orbits the sun.

Maybe the so-called miracle of science happened only to us: I don't really know, but if I've learned anything from my lengthy immersion in the ethnosciences, it is that if you want to speak with any confidence about what a people thought, you'd best study *their* culture, or hang out with those who have. The native astral myths my fellow investigators encountered in the historical record served only to embellish their findings. Let me illustrate with the two most well-known Native American myths about eclipses:

When this came to pass he [the sun] turned red; he became restless and troubled; he faltered and became yellow. Then there were a tumult and disorder. All were disquieted, unnerved, frightened. Then there was weeping. The common folk, raised a cup, lifting their voices, making a great din, calling out, shrieking. There was shouting everywhere. People of light complexion were slain [as sacrifices]; captives were killed. All offered their blood, they drew straws through the lobes of their ears, which had been pierced. And in all the temples there was the singing of fitting chants; there was an uproar; there were war cries. It was thus said: "If the eclipse of the sun is complete; it will be dark forever! The demons of darkness will come down; they will eat men!"[1]

Bernardino de Sahagun, chronicler of the Aztecs, was brought over to Mexico in the aftermath of the Spanish conquest to aid in Christianizing the native population. In his commentary he focuses on the sense of fear felt and the noisy reaction made by his informants on witnessing the solar eclipse of 1531. Garcilaso de la Vega, Sahagun's counterpart among the Inca of Peru, gives a similar account of the agitated behavior of eclipse watchers:

When there was a solar eclipse, they said that the sun was angry at some offence committed against him, since his face appeared disturbed like that of an angry man! And they foretold, as astrologers do, the approach of some great punishment . . . they were seized with fear and sounded trumpets, bugles, horns, and all the instruments they could find for making a noise. They tied up their dogs, large and small, and beat them with many blows and made them howl and call the moon back, for according to a certain fable they told, they thought that the moon was fond of dogs in return for a service they had done her and that if she heard them cry she would be sorry for them and awake from sleep caused by her sickness.[2]

DESOLATION des PERUVIENS pendant L'ECLIPSE de LUNE

Imagined Inca eclipse worshippers beating their dogs to call back the moon. (Photo courtesy of The Philadelphia Print Shop, Ltd., philaprintshop.com)

As far as the European intruders were concerned, these stories were manifestations of the idolatrous worship of a heathen people, whose superstitions they would need to learn if they were to successfully subdue their ideology. As one of the more zealous chroniclers put it: "Paganism must be torn up by the roots from the hearts of these frail people!"[3]

We may not feel so threatened by these unusual people as to want to tear their hearts out, but we still harbor suspicions about them. Wedded to the idea of progress, we link what others believe with customs handed down to us by our own ancestors, like carrying a rabbit's foot for good luck, or saying "God bless you" after a sneeze, or knocking on wood. These amusing appendages of past ways of thinking that have managed to survive cultural change fit the definition of the term superstition, which means literally "to stand above cause." Today's dictionaries associate these ancient superstitions with beliefs in which religious veneration is shown where none, based on faith, is deserved. In the absence of causal evidence, one worships false gods through fear, as the chroniclers implied. Little has changed.

But who decides what's rational, what unknown warrants being feared, and what lies within the realm of plausible explanation? We rely on common sense, a culturally shared belief in understanding the world in a way we've all grown accustomed to and, out of habit, adopt as part of the way we live. I think what makes superstition a pejorative term is the implied, deep-seated connection with belief in a direct connection between people and the powers of nature: magic. When these powers are deemed evil, negative words like "witchcraft," "omen," and "cult" find their way into the definition. This is why such terms come to be characterized as deviant forms of religious worship. Common sense, for most of us, is based on the firm conviction that the material world has no direct link with or power over the human condition. The very thought of someone who would actually believe in the efficacy of material objects, rabbit's feet for luck or wood for knocking, in dealing with human affairs bothers our rationally cultivated minds.

By helping synthesize a new cultural astronomy perspective, I've learned that searching out the roots of what the members of other cultures say they feel and how they react when they witness a total eclipse of the sun doesn't tell us much about the science of eclipses, but it does open up a doorway to knowledge about people and their beliefs. Let me illustrate by putting contemporary African Lozi and Lamba, Maya, Hindu, Inuit (Canadian), and Indonesian eclipse stories in context.

Cultural anthropologists pay particular attention to variations in behavior among different societies. When it comes to human reactions to a total eclipse of the sun, professional people watchers are in no better position than the animal watchers to collect data on their subjects. Anthropologists can confront, interview, and survey the living descendants of people who witnessed eclipses. Many reported beliefs have obviously been influenced by European Christian doctrine. Anthropologists can further explore continuities and disjunctions between cultural data from past and present; for example, they look for eclipse lore embedded in myths passed down by oral traditions that offer a social context for behavior that has been passed off as a product of superstition. Unfortunately, the effects of totality on people are rarely mentioned in books about eclipses.

Of all the reactions to eclipses I have heard about, making noise and biting, eating, or swallowing stand out. Notice the noise-making in both of the Spanish chroniclers' eclipse narratives. We all make noise for different reasons: to attract attention, to warn a companion of imminent danger, or to scare away an animal we encounter in the woods. There is noise-making too among contemporary cultures; for example, people banged pans together during the 1973 eclipse in South Sudan, and Zulu clapping and wailing lamentations can be heard from afar when the sun is eclipsed. They say the people must awaken the solar deity from his lethargy, for his inattentiveness fore-tells a great calamity. Like the god, the people become sluggish; no man slept with his wife, women stopped brewing beer, men quit hunting and slaughtering, and cows went unmilked.[4] Noise-making is also pervasive in eclipse watching in the contemporary Maya world. Ants, or alternatively jaguars, are devouring the sun, so we need to make as much noise as possible, even pinch our dogs to make them howl. Some Maya say eclipses are renewals of an old cosmic sibling rivalry. The two luminaries are fighting over lies the moon once told the sun about how people on earth have been behaving. Noise made during an eclipse is intended to get the sun's attention. We need to convince the sun that these stories aren't true.[5]

Other tribal cultures offer a sexual interpretation for the noise-making during sun-moon interactions. The abnormal hookup of the two luminaries invokes the theme of incest. The Great Plains Arapaho interpret the sun and moon changing places in the sky during an eclipse as a gender role reversal; in other words, the normal gender relationship, the sun as male and the moon as female, is violated when the two join together in eclipse. One mythological interpretation of Arapaho marriage is seen "as a monthly honeymoon interrupted at each dark moon by menstruation."[6] In effect, a woman's monthly time division between sex with and seclusion from her husband is coded celestially as an alternation between the sun and the moon.

Anthropologists have uncovered a connection between incest and making noise in a number of rites associated with natural phenome-na, especially eclipses, darkness, and storms, and the flow of blood and unruly behavior in general.[7] The link between noisy rebellion, cooking, and incest appears in one account of a Brazilian myth (a version that has much in common compared with myths across Native America): A man had engaged in sex with his sister every night, but never revealed his identity. She stained his face with *genipa* (a relative of the gardenia) juice to identify him publicly. He went to live in the sky, and there he became the moon. (Ever notice those spots on his face?) The sister followed him into the sky; they quarreled, and she fell to earth, making a loud noise. Another brother learned of this and shot arrows at the moon. People below were spattered with his blood. When they attempted to smear it away, they acquired the colors of their dress plumage. They ceased cooking out of fear that the descending blood would pollute their food. Thoughts of lunar blood are triggered by the eclipse.

Dynastic power and the meaning of history are also linked to the cosmos among the contemporary Lozi, an ethnic group comprising half a million people who live along the Zambezi River in Zambia. Continuity in rulership and balance in relation to its citizens are expressed in rites based on observing the sun and the moon. Elders say that at the beginning of time their sun god, Nyambé, married Nesilele, the moon. Together they came down from heaven and gave birth to the first in the line of kings destined to govern the people of the region and to safeguard the secret knowledge of the mystery of the heavens. Every year during the seasonal flooding of the river timed by the first full moon following, they celebrate the Kuomboka pageant, recognizing the continuation of dynastic power. Kuomboka means "get out of the water"; its significance is to warn both royalty and subjects to vacate their houses and get to high ground. Royal drums give the signal as the illuminated lunar disk rises high in the sky. In the ceremony, the Lozi kings lead their people away from danger by boarding a black-and-white-striped river barge, black for dark storm clouds and white for the cool drops of rain that emanate from them. African art historian Karen Milbourne thinks that the black and white also stand, respectively, for death balanced by renewal and ancestorhood; both are lunar symbols.[8] Kuomboka climaxes just before the arrival of the next new moon, when the barge reaches dry land. Amid loud drumming and a cheering crowd, the king gestures to the setting sun while the onlookers clap their hands, reinforcing the connection between the two sky deities. Meanwhile, his attendants move around the drums in a circle, the way the sun and moon cycle around the sky.

The Kuomboka ritual appears to be designed to show that the king of the Lozi acquires his legitimacy to rule from his ancestors and he must continually renew it in ceremonies timed by the moon; however, as Milbourne points out, it is the sun who influences the way the king interacts with the people. Hand clapping isn't done out of fear; it's more like applause, part of the Lozi code of conduct for recognizing the sun as the force in the sky that seals the bond between the ruler and his people. A nineteenth-century account of this noisemaking tells of one company of men clapping in unison before their king: "and before taking their places [they] raised their hands above their heads and shouted the royal salutation, *'Yo sho, yo sho, yo sho!'*"[9] Then, as they continue clapping, celebrants bow to the earth three times. The ceremonial clapping unites heaven and earth, restores order, achieves balance, and shows respect for their leader. The ritual continues today.

Once we abandon the acquired habit of interpreting things literally, we begin to see that the goals of acquiring knowledge from the sky differ from culture to culture. The Lamba of Western Zambia, northern Togo, and Bénin use kinship terms to describe relations between interactions of sky objects. Their stability and endurance offer people a permanent model to address questions about how to live. The moon becomes the nephew of the sun, and various morning and evening star Venus pairings become husband and wife, mother and child, brother and brother, sister and sister, and the all-important mother's brother and sister's son. God is the father-in-law of Thunder. When he claps his hands, it's a sign of respect for his father.

Sun rules day, moon rules night; sun gold, moon silver; sun male, moon female. These are just a few pairings that highlight the cross-cultural solunar dualism. One of the oppositions of the sun and moon best seen from the tropical latitudes of sub-Saharan Africa has to do with how high they arc across the sky. Most northern hemisphere skywatchers are aware that winter full moons ride high in the sky in the south, while their seasonal counterparts around the summer solstice pass low in the south. At latitudes close to the equator, the extremes carry competitors of luminosity across the zenith. When a full moon passes overhead, its light invites comparison with the sun. (Have you ever heard of being moonstruck: dazzled, crazed, abnormally affected by the light of the moon?)

Sun/moon oppositions also correspond with the rainy/dry seasons in Zambia. Their summer is characterized by a low full moon and rain, which forms mud. The high-flying (pure) sun symbolizes clean chores, while the low-slung "filthy" moon refers to dirty chores. They say that the moon has now been chased away or "besmirched" by his brother, the sun.[10] On occasion he can even be captured or devoured, or in turn devour his partner: an eclipse.

The moon biting the sun is another frequently reported interpretation of an eclipse. "Do people bite or eat one another in your country?" a Chamula Maya descendant asked anthropologist Gary Gossen during an interview. "Of course not," came the obvious answer. "Do people bite and eat one another here?" Not now, was the response, "but the first people did."[11] As Gossen continued his line of questioning, he began to realize that, as far as the Maya were concerned, Americans lived on the outer fringes of the world and what lay far away in spatial distance also resided in ancient time. Even though the Chamula had civilized themselves by creating social rules eliminating cannibalism and infanticide from contemporary space-time, they believed that such deviant behavior might still occur at the outer limits of the universe; that's why the informant posed his question.

If the competing sources of good and evil yet reside in the universe where the sun and moon also dwell, the eclipse becomes a reminder that the social order is always in danger of getting out of balance. The ideal society the Chamula envisage, free of the chaotic social behavior of ancient barbarism, has yet to be realized. Is this so different from the Christian celestial metaphor of the permanent darkness that will prevail upon sinners at the time of the final judgment? Far from something to be feared, the eclipse serves as a platform for contemplation and discourse about the forces of nature for the contemporary Maya, just as it does for us: "A startling nearness of the gigantic forces of nature and their inconceivable operation seems to have been established. Personalities and towns and cities, and hates and jealousies, and even mundane hopes, grow very small and far away."[12]

In parts of rural India, the solar eclipse emerges as a mediating force between tradition and progress. According to Hindu mythology, a solar eclipse occurs when the demon Rahu swallows the sun. Like the Maya, they say that ancient times were filled with chaos as the demons and the gods fought over who would rule the world. Out of the churning ocean, Lord Vishnu, protector of the universe, created a pot of Amrit, the all-powerful elixir of immortality, for the contestants to consume. He disguised himself by changing into a woman before distributing the magical brew to both sides, who were seated in orderly rows. But Vishnu stacked the deck. He tricked the demons by offering them an imitation liquid. Seated next to the sun and moon deities, one clever demon realized he was being duped. He grabbed a goblet of Amrit and began to drink it. Vishnu suddenly assumed his real form and beheaded the demon on the spot, so that the elixir could not take effect. The demon's head, called Rahu, swallowed the sun out of revenge. Whenever he gets the chance, he repeats his act of consuming the sun.

A British visitor who came to India to witness the August 18, 1868, eclipse and heard of the Rahu eclipse myth commented: "European science has as yet produced but little effect upon the minds of the superstitious masses of India. Of the many millions who witnessed the eclipse of the 18th of August last there were comparatively few who

According to Hindu mythology, a solar eclipse occurs when the demon Rahu swallows the sun. (Wikimedia Commons)

did not verily believe that it was caused by the dragon Rahu in his endeavor to swallow up the Lord of the Day."[13] He added: "The pious Hindu, before the eclipse comes on, takes a torch, and begins to search his house and carefully removes all cooked food, and all water for drinking purposes. Such food and water, by the eclipse, incur *Grahama seshah,* that is, uncleanliness, and are rendered unfit for use. Some, with less scruples of conscience, declare that the food may be preserved by placing on it *dharba* or *Kusa grass.*"[14] Reading the first part of that statement gave me a sense of the imperialist mentality of an outsider impatient over failure to colonize the Hindu mind. But the second part connects eclipse watching with a deeper meaning worth exploring in Hindu culture: subsistence.

More than a century later, a group of Indian anthropologists conducted a rare study of the response of diverse Hindu people to a solar eclipse, one that took place on October 24, 1995, and was total in the Ganges area of West Bengal. They were interested specifically in the impact of the electronic media on traditional beliefs, like the Rahu story. In the urban societies they worked with, the anthropologists found that women in particular mentioned taboos on cooking, eating, and drinking, especially swallowing. One woman told an interviewer that she intended to avoid cooking; another said she planned to throw away the supply of drinking water in the house and resist members of her family swallowing anything during the eclipse. Her educated, more well-traveled young son said that he would obey his mother's wishes, not because he believed harm would come to him by swallowing anything, but to avoid hurting his mother's feelings or clouding family sentiment.

Another mother told of her intention to take a post-eclipse bath in the Ganges to remove any residual negative effects of Rahu. When questioned about these beliefs, her more educated daughter insisted that this is not so much a superstition as part of tradition: "Why should [I] indulge in violating the custom and risk the well-being of family members?"[15] Why was she conflicted? Just as some of us might withhold breaking with our family's cultural or religious traditions despite a more independent or secular personal view, younger family members pointed out that although they were perfectly well aware of the scientific explanation of eclipses, they opted for the cohesive power of following family customs. Aren't many of us conflicted as we try to cope with the implications of scientific discovery in a vast material universe with which we've lost intimacy?

Among rural Hindu populations, the anthropologists discovered eclipse reactions even more deeply embedded in tradition. Members of one tribe told a diluted version of the story of Rahu swallowing the sun: the moon god became angry with the sun god, so he swallowed him. One variation has it that the moon borrowed some rice (standard currency in ancient times) from the sun. Unable to pay it back, the moon hid behind the sun out of remorse; that's what eclipses are about.

Another spin on the rice-borrowing story finds the sun hiding behind the moon because he thinks the latter has returned for seconds. In one ritual that emanated from the rice debt account, people put some rice grain along with a coin (modern currency) in a packet on the thatched roof of the house. After the eclipse, the rice was kept for use in other sacred rituals. Following along the line of some of the principles of magical behavior I mentioned earlier, like knocking wood, some Hindus expose their agricultural and hunting implements, along with a piece of iron bar, to the eclipse by placing them on the roof. The bar becomes part of the material used to fabricate other household articles. If there's a newborn in the house, they place some of the exposed rice grain next to the child and later disperse it outside the house during the night to ward off any remaining evil shadows. Some nontribal people interpret the rice borrower to be a member of a lower caste. In effect they transfer the relationship between the eclipse and food spoilage to the idea of purity and pollution.

Wherever I look, I find deep-seated social and religious beliefs at the base of transcultural eclipse watching. In the northeast Canadian Inuit province, young Mark Ijjangiaq's eclipse story typifies the educated younger generation's loss of childhood fear. Though he told the story with some amusement, he still participated fully in the ritual:

> One afternoon when I was inland with my family for the summer there was an eclipse of the Sun. I was nine years old at the time and indeed I was horrified by the experience. It had the same effect as when one is wearing sunglasses. The Sun had the same characteristics as the Moon when it is just coming into its first or last quarter. My mother was still alive at the time. We had a place we drew water from—a small stream, very close to our tent. After the eclipse was over my mother asked me to go and fetch water with a small pail. But I did not want to go for I was afraid that once I went a little distance [from the tent] there might be another eclipse while I was all alone.[16]

I wonder what Mark really feared. Why did he mention his mother? We can't know with any certainty, but the threat of being alone at least confirms the sense of assurance his mother's proximity must have provided during their viewing of the eclipse. She could not vanish as long as he was next to her, and his personal hold over reality could be confirmed in her eyes.

The Inuit say that during an eclipse all animals and fish disappear. To get them back, hunters and fishermen gather samples of every kind of creature they consume and place them in a large sack. Carrying their burden, they circuit the periphery of the village, acting out the direction of movement of the sun. Back in town, they empty the sack and distribute bits of each kind of flesh to residents to eat. The central theme is, again, subsistence. Mutual respect between hunter and prey is an important part of Inuit culture. Inuit tell inquiring anthropologists the creatures remind them that they need their attention and the only way successful hunting can be resumed after the eclipse is for the men to perform this rite so that they can return safely. The eclipse becomes the medium through which man engages beast.

Following the February 21, 1877, eclipse, a sea captain reported that the eclipse took place in the afternoon, while a number of Eskimo happened to be on board his ship: "They appeared much alarmed, and with one accord hurried out of the ship. Before they were all on the ice a brisk squall came on, and added not a little to their terrors. Okotook ran wildly about under the stern, gesticulating and screaming at the sun, while others gazed on it in silence and dread. The corporal of the marines found two of the natives lying prostrate with their faces to the ice quite panic-struck. We learned that the eclipse was called *shiek-e-nek* (the sun) *tooni-lik-pa.*"[17] And in East Greenland: "When an eclipse of the Moon takes place, they attribute it to the Moon's going into their houses, and peeping into every nook and corner in search of skins and eatables, and on such occasions accordingly, they conceal all they can, and make as much noise as possible, in order to frighten away their unbidden guest."[18]

In the first account, the captain, like the Spanish chroniclers or the British visitor to India, makes no effort to explain why the natives reacted that way. Did he take them for fools? And what of the second story? Surely it isn't a literal interpretation of what the native thinks will happen? Anthropologist John MacDonald tells us that "in order to avert or reduce the phenomenon's expected ill effects, eclipses of both sun and moon often required specific actions to be taken, or taboos to be observed, on the part of those experiencing them: He [the Moon] rejoices when the women die, and the sun in her revenge has her joy in the death of men; therefore the men keep within doors at the eclipse of the sun, and the women at the eclipse of the moon."[19] Family ties enter the picture when the sun and moon are personified; the moon acts this way because he is ashamed of how he mistreated his sister. In another account the brother pursues her across the sky; when he catches up to her and hugs her, it causes the eclipse.[20]

The Suki of Papua New Guinea told Dutch anthropologists in the 1960s that eclipses happen because people's souls leave their bodies and throw themselves at the sun or the moon. If a soul fails to return to its body, the possessor will die. Those responsible for solar eclipses are members of the cassowary, or large bird moiety, one of the paired kinship groups into which many cultures around the world divide themselves. The cassowary moiety is associated with water, the rainy season, darkness, and the moon, while the pig moiety relates to land, the dry season, and bright days; they tend to lunar eclipses.

Eclipses happen when members of one moiety cross over to the other, thus violating the natural order. After considerable difficulty getting anyone to talk about eclipses, the anthropologists finally spoke to two informants who had witnessed such events over the years and had been surprised by them. In each instance an appointed clansman from the tribe responsible climbed a tree and asked the sky gods what it all meant. In one case, four souls from sick people in the opposing clan were covering the sun. In another, a woman whose husband had abused her was named. She died shortly after the eclipse. In a third incident, the clansman went to the top of the tree and shouted, "Nagaia Namagwaria, Gwauia, what is the matter?" Came the omen: "The man here is Tutie, we tried to send him back, but he did not obey us. We have cut his hair so that he will soon die."[21] When Tutie learned he'd been fingered, he began to cry. He too, they said, died shortly after the eclipse.[22] That the Suki seem to go to great lengths to conceal their eclipse traditions from outsiders suggests a deep commitment to their beliefs.

Making noise, eating, biting, swallowing: what puzzles us about these eclipse stories is the apparent conflict between myth and cognition. Why should an animate sun and moon serve as the action figures in the reenactment of a narrative of life? How is it that these mental speculations continue to coexist alongside rational thought, so clearly evident in the Hindu eclipse reactions? I think the common thread that knits all humanity together is deeply woven in the human desire to embrace intangible natural phenomena, for us black holes in an infinite universe, for others a total eclipse of the sun, by trying to make sense of what we see with what we experience in the tangible world. For example, astronomers speak of births and deaths of stars and black holes consuming one another and belching after a cosmic lunch. Only by lending familiarity to the unfamiliar can we hope to find meaning in it. The eclipse stories I've narrated in this chapter are examples of the way people project their social interactions into the world of nature, especially where the status quo is violated; for example, by threats to sustenance, breaching family ties, and political instability. People create stories about the world around them as a way of transporting that unpredictable realm closer to themselves. By discovering society mirrored in the cosmos, they humanize themselves.

So it isn't the eclipse that matters; it's the action triggered by what happens to the one who sees it. For the Maya, daytime darkness is a reminder of challenges to the social order and the need to engage a discourse to preserve it. The Hindu sun watcher recalls the obligation to debt payment and the cohesive power of family custom in the face of techno-modernity, while the Inuit, in a subsistence-challenged environment, perform a rite to sustain harmony between hunter and hunted. In all societies outside the modern West, the sun and the moon are *not* members of a world apart, a world of matter devoid of spirit, as science teaches us. The celestial players involved in the cosmic drama of eclipse offer a kind of dualistic imagery that reflects complementary social concepts in everyday life: male and female, pure and impure, good and evil, night and day. Far from amusing relics, eclipse stories from around the world are examples for the active mind to engage and contemplate. They become powerful agents in setting up a dialogue between people and the sky about the meaning of human existence. They restore a lost moral component to the art of skywatching, and they inspire us to pay closer attention to human diversity.

AFTERWORD
A Confession

I waited two or three moments: then looked up; he was standing there petrified. With a common impulse the multitude rose slowly up and stared into the sky. I followed their eyes; as sure as guns, there was my eclipse beginning! The life went boiling through my veins; I was a new man! The rim of black spread slowly into the sun's disk, my heart beat higher and higher, and still the assemblage and the priest stared into the sky, motionless. I knew that this gaze would be turned upon me, next. When it was, I was ready. I was in one of the most grand attitudes I ever struck, with my arm stretched up pointing to the sun. It was a noble effect.

—*Mark Twain,* A Connecticut Yankee in
King Arthur's Court, *1890*

The dark of a total eclipse is nothing like night, and it isn't like twilight or dusk, either. This kind of darkness comes without a sunset, and its shades and tones evolve over a vastly speeded-up time scale. Though I've seen it happen eight times, still I struggle to relate exactly what it is—this absence of light. I know I always feel enlightened, "noble," as Mark Twain says, when the cheer goes up at first contact. Is it because I know we can predict when it will happen to the nearest second, the same feeling that caused a cadre of physicists to cheer when they finally detected gravitational waves, or the engineers in mission control to applaud when astronauts first landed on the moon? At the same time, my senses are overwhelmed by the sudden appearance of the reds and lilacs, the peach, crimson, and topaz colors of the prominences and the corona. Why should I experience any uneasiness in anticipation of something I can foretell with absolute certainty?

I am the mediator, the one who picked out those choice epigraphs that head each chapter in my aim to acquaint you with signal aspects that have attended the eclipse experience through the ages and across cultures, hoping to prompt your rational and emotional as well as your scientific and religious expectations. Still, from beginning to end, I struggle to find a conceptual reference point around which to build my narrative.

Expressing extraordinary sensory impressions to the nonparticipant is really a problem of language.[1] People struggle to describe firsthand experiences of tornadoes: "it was like a train . . . there was this roaring sound like I never heard"; or a total eclipse of the sun: "I never saw anything like it . . . it was a miracle." If nothing like it ever happened to you, where do you acquire the appropriate vocabulary to tell about it? How can you cast an entirely novel, overwhelming experience in a familiar framework?

Fear and superstition intrude on us to varying degrees once the lights go out. We are surrounded by that strange darkness, and darkness itself has long been associated with uncertainty in our minds. It's the discontinuity, the disruption that's so bothersome—even if it's planned. The whole luminous affair is a mockery of the normal oscillation between light and dark that happens every day. Then there's the aesthetic component, the feeling of something sublime. Because the darkness of the eclipse doesn't correspond to anything we experience in lived time, some eclipse watchers choose death as a metaphor, for want of a better analogy.[2]

So, after much thought and with apologies for any failings, these are the words I ended up choosing in my attempt to join together the objective (to understand) and the subjective (to feel) of what it's like to witness a total eclipse. Now you must see for yourself.

APPENDIX

A Brief Chronology of Solar Eclipses

Date	Location	Remarks
BCE		
October 22, 2137 (annular)	China	Chinese astronomers fail to predict an eclipse(?)
May 13, 1375	Middle East to Asia	Earliest record of a solar eclipse
April 16, 1178	Central Mediterranean (south to north)	Eclipse mentioned in the *Odyssey*
May 28, 585	Southern Europe, Greece, Asia Minor	Thales foretells an eclipse during the war between the Lydians and the Medes
August 3, 431 (annular)	Western Black Sea, central Asia Minor	Thucydides predicts an eclipse during the Peloponnesian War
CE		
November 24, 29	Asia Minor, Syria, Arabia	Eclipse occurring at the time of the Crucifixion of Jesus
November 24, 569	Horn of Africa	Eclipse at the time of the birth of Muhammad
January 27, 632	Sudan, Yemen, India	Eclipse at the time of the death of the Son of Muhammad
November 11, 923	Middle East to India	Arab astronomers make precise observations of eclipse

APPENDIX

Date	Location	Remarks
August 2, 1133	Northern England, Central Europe, Greece, Arabia	Eclipse allegedly presages death of Henry I; alternatively August 18, 909, or June 28, 145
August 22, 1142*	United States: Great Lakes and Northeast	Iroquois Federation Foundation eclipse
April 13, 1325*	Central Mexico, United States: Florida	Aztec Foundation of Tenochtítlan
July 29, 1478*	United States: Eastern	Last total eclipse in New York City prior to 1925
August 8, 1496	Northeast to southeast Mexico	Total eclipse pictorially recorded at Tenochtítlan by the Aztecs
July 20, 1506*	Central United States	Total eclipse at site of colonial Jamestown, United States, one year before foundation of British colony
October 12, 1605	Pyrenees, Mediterranean, Asia Minor	First scientific description of corona
May 30, 1612	Northern Scandinavia	First eclipse viewed through a telescope (partial phases)
October 23, 1623*	United States: North Carolina	
March 8, 1625*	United States: Central Florida	
March 29, 1652	England, Ireland	Astronomy versus astrology in eclipse debate
August 22, 1672*	United States: western and south central Maine	
May 3, 1715	Southern England, Scandinavia	Edmond Halley enlists public support to document eclipse
October 4, 1717*	United States: Carolinas	
May 11, 1724	Ireland, southern England	Edmond Halley's second eclipse

Date	Location	Remarks
May 12, 1733	Scandinavia	Prominences (flames) first observed at edge of disk
May 13, 1752*	United States: southern Florida	
January 24, 1778*	United States: New Orleans, Louisiana, to southern Virginia	David Rittenhouse is first American astronomer to observe eclipse
October 27, 1780*	Maine	Failed Penobscot Bay expedition
January 16, 1806*	United States, transcontinental: Southwest to central New England	
September 17, 1811*	United States: Northeast	Annular; Thomas Jefferson's eclipse
February 12, 1831*	United States: Mississippi to Virginia	Annular; Nat Turner's eclipse
November 30, 1834*	United States: south of Memphis, Tennessee, to Savannah, Georgia	
May 15, 1836	Great Britain, northern Europe	Baily's beads first noted
July 8, 1842	Central Europe, Central Asia, China	Corona and prominences recognized as part of solar atmosphere
July 28, 1851	Scandinavia, Germany, west central Asia	First eclipse photo (daguerreotype)
July 18, 1860	Spain, Algeria	First eclipse photo (wet plate)
August 7, 1869*	United States: northern, central, Carolinas	Discovery of coronium in solar spectrum
July 29, 1878*	United States: Northwest, Rocky Mountains, Texas, Louisiana	Pike's Peak eclipse
January 1, 1889*	United States: Northern California to North Dakota	
May 28, 1900*	United States: Louisiana to southern Virginia	

Date	Location	Remarks
August 30, 1905	Spain, southern Mediterranean, Egypt, Arabia	
June 17, 1909	Greenland	Triple saros great-grandparent of June 20, 1963, eclipse
February 3, 1916*	Northern South America	Triple saros great-grandparent of March 7, 1970, eclipse
June 8, 1918*	United States: Washington, Oregon, central Florida	Most recent transcontinental total solar eclipse in the United States
May 29, 1919	Brazil, south central Africa	Einstein's eclipse
January 24, 1925*	United States: Great Lakes to New York City and Long Island	
August 31, 1932*	United States: Maine, central New England	Biologists' eclipse
July 9, 1945*	United States: central Idaho, Montana	
June 30, 1954*	United States: Kansas, Nebraska, Minnesota, Michigan	
June 20, 1955	South Asia, Indonesia	Longest duration eclipse of twentieth century (7^m08^s)
October 2, 1959*	United States: New Hampshire, northeast Massachusetts	
February 5, 1962	Indonesia, South Pacific	"End of the world" eclipse (February 4) in Los Angeles, California
July 20, 1963*	United States: Maine	Great Maine eclipse
March 7, 1970*	United States: northern Florida, coastal South Carolina, North Carolina, Virginia, Nova Scotia	Triple saros great-grandparent of April 8, 2024, eclipse
July 10, 1972	Northern Canada, North Atlantic	First major astronomy education eclipse cruise

Date	Location	Remarks
June 30, 1973	Central Africa	Long duration eclipse (7^m04^s); parent of July 11, 1991, eclipse
October 12, 1977	Mid-Pacific, Columbia, Venezuela	A pair of Columbus Day eclipse cruises rendezvous
February 26, 1979*	United States: Washington, Oregon, Idaho, Montana, North Dakota	Last eclipse to touch mainland United States
July 11, 1991	Mexico, United States: Hawaii	Long duration eclipse (6^m51^s)
November 3, 1994*	South America, South Atlantic, United States: East Coast	
October 24, 1995	India to southeast Asia	First organized study of eclipse responses by anthropologists
August 11, 1999	South England, central Europe, Turkey, Iran, India	Saros parent of 2017 eclipse, "last eclipse of the millennium"
June 21, 2001	South central Africa	Ravers' solstice eclipse
December 4, 2002	South Africa to southern Australia	Outback eclipse
March 29, 2006	North Africa, Turkey, Central Asia	
July 22, 2009	India, China	Longest duration eclipse of the twenty-first century (6^m39^s)

Future Eclipses

August 21, 2017*	United States: Oregon to South Carolina	
April 8, 2024*	Mexico, United States: Texas, Midwest, Great Lakes	
August 12, 2026	North Atlantic, Spain	
August 2, 2027	North coast of Africa, Egypt	
July 22, 2028	Australia	

Date	Location	Remarks
November 25, 2030	Australia	
July 13, 2037	Australia	
December 26, 2038	Australia	
August 12, 2045*	United States: Northern California to Florida	
March 30, 2052*	United States: southern Louisiana to southern South Carolina	
September 23, 2071*	United States: Oregon, Rocky Mountains, Texas, Louisiana	Triple saros great-grandchild of August 21, 2017, eclipse
May 11, 2078*	United States: New Orleans to southern Virginia	Near repeat of 2052 eclipse; Triple saros great-grandchild of April 8, 2024, eclipse
May 1, 2079*	United States: Pennsylvania, New Jersey, New York City, Cape Cod	Next total eclipse in New York City
September 3, 2081	Central and eastern Europe, southern England	
September 23, 2090	Central and eastern Europe, southern England	
September 14, 2099*	United States: North Dakota to southern Virginia	

Source: Adapted from NASA, "Solar Eclipses of Historical Interest," available at eclipse.gsfc.nasa.gov/SEhistory/SEhistory.html.
*United States mainland eclipse.

Notes

Chapter 1. Colossal Celestial Spectacles

Epigraph: William Shakespeare *King Lear,* Act 1, Scene 2.

1. When it crossed the earth's orbit, our planet was nowhere in the vicinity.

2. See my *The End of Time: The Maya Mystery of 2012* (Boulder: University Press of Colorado, 2009) and *Apocalyptic Anxiety: Religion, Science, and America's Obsession with the End of the World* (Boulder: University Press of Colorado, 2016).

3. "History of the Leonid Shower," available at leonid.arc.nasa.gov/history.html.

4. Ibid.

5. John MacDonald, *The Arctic Sky: Inuit Astronomy, Lore, and Legend* (Toronto: Royal Ontario Museum, 1999), p. 149.

Chapter 2. Watching People Watching Eclipses

Epigraph: Archilochus, quoted in F. Richard Stephenson, *Historical Eclipses and Earth's Rotation* (Cambridge: Cambridge University Press, 1997), p. 338.

1. Eileen Blass, "The Thrill of the Chase, the Fury of the Storm," *USA Today,* April 8, 2014.

2. "Total Solar Eclipse 2017," available at Eclipse.org.

3. "Sun Was Eclipsed Behind Cloud Veil," *New York Times,* August 22, 1914, p. 13.

4. Note that we are using the standard notation h = hours, m = minutes, s = seconds.

5. Lisa Grossman, "Longest Eclipse Ever: Airplane Chases the Moon's Shadow," Wired.com, July 22, 2010.

6. Antonio Regalado, "The Other Eclipse People Are Flocking to See," *Wall Street Journal,* July 6, 2010.

7. "The Eclipse Chasers," *Daily Mail,* March 19, 2015.

8. Regalado, "The Other Eclipse."

9. Kate Russo, *Total Addiction: The Life of an Eclipse Chaser* (New York: Springer, 2012).

10. Duncan Steel, *Eclipse: The Celestial Phenomenon That Changed the Course of History* (Washington, DC: Joseph Henry Press, 2001), p. 227.

11. Gregory Sams, "Solipse 2001 Party—Review," available at gregorysams.com/solipse-2001.html.

12. James Norman, "10,000 Revelers Expected at Eclipse Rave," TheAge.com, November 30, 2002.

13. Graham St. John, ed., *Rave Culture and Religion* (London: Routledge, 2004).

14. George Smith and William Makepeace Thackeray, "Great Solar Eclipses," *The Cornhill Magazine* 18 (1868): 161–63.

15. Bertrand Russell, *Why I Am Not a Christian, and Other Essays on Religion and Other Related Subjects,* ed. Paul Edwards (New York: Simon and Schuster, 1957), p. 24.

16. James W. Jones, *Can Science Explain Religion?: The Cognitive Science Debate* (Oxford: Oxford University Press, 2016).

17. All biblical quotations from *The New Oxford Annotated Bible with the Apocrypha,* ed. Herbert May and Bruce Metzger (New York: Oxford University Press, 1977).

18. One of them even made the top fifty list on Amazon.com: John Hagee, *Four Blood Moons: Something Is About to Change* (Franklin, TN: Worthy, 2013).

19. Robert Richardson, "The 'End of the World' at the Griffith Observatory," *Griffith Observer* 26, no. 5 (1962): 62–65 (quotes on p. 62).

20. "Meaning and Effect of a Solar or Lunar Eclipse," n.d., Spiritual Research Foundation, available at SpiritualResearchFoundation.org. For more on the Hindu practice of not consuming food during eclipses, see Chapter 17.

21. Sandra and David Mosley, "Solar Eclipses," n.d. available at ZodiacArts.com.

22. Anthony Aveni, *Conversing with the Planets: How Science and Myth Invented the Cosmos* (New York: Times Books, 1993); Anthony Aveni, *Behind the Crystal Ball: Magic, Science, and the Occult from Antiquity Through the New Age* (New York: Times Books, 1996).

23. "What Do Americans Believe?," The Harris Poll, available at www.theharris-poll.com/in-the-news/harris-polls/Americas-Belief-in-God.html; "Paranormal Beliefs Come (Super)Naturally to Some," November 1, 2005, Gallup poll, available at www.gallup.com.

24. Chris Mooney, "More and More Americans Think Astrology Is Science," *Mother Jones,* February 11, 2014. For data going back into the twentieth century, see my *Behind the Crystal Ball,* pp. 246–48.

25. Henri Frankfort, *Ancient Egyptian Religion: An Interpretation* (New York: Dover, 1948), p. 29.

26. "Voyage to Darkness—In Search of Eclipses," Nauticom.net.

27. Bill Kramer, "Report and Images of the 1973 Total Solar Eclipse," available at Eclipse-Chasers.com.

Chapter 3. What You See and Why You See It

Epigraph: Pindar, *Ninth Paean,* addressed to the Thebans. Quoted in F. Richard Stephenson, *Historical Eclipses and Earth's Rotation* (Cambridge: Cambridge University Press, 1997), p. 344.

1. "Anatomy of a Solar Eclipse," Naturphilosophie.co.uk.

2. A competing, less favored, theory is that the bands are caused by the supersonic speed of the moon's shadow, which produces a shockwave that arrives as totality approaches.

3. Rebecca Joslin, *Chasing Eclipses: The Total Eclipses of 1905, 1914, and 1925* (Boston: Walton Advertising and Printing, 1929), pp. 14–15.

4. Francis Baily, *Monthly Notices of the Royal Astronomical Society* 4 (1836): 15.

5. See Chapter 14 for details about how the Diamond Ring was named.

6. Mabel Loomis Todd, *Total Eclipses of the Sun,* rev. ed. (Boston: Little Brown, 1900), p. 21.

7. Mark Littman, Fred Espenak, and Ken Willcox, *Totality: Eclipses of the Sun,* 3rd ed. (Oxford: Oxford University Press, 2009), p. 136.

8. Joslin, *Chasing Eclipses,* p. 127.

9. Positive and negative health benefits of fluorescent lighting have been reported, but the most thorough studies conclude no dramatic effects are apparent. See Shelley McColl and Jennifer Veitch, "Full Spectrum Fluorescent Lighting: A Review of Its Effects on Physiology and Health," *Psychological Medicine* 31, no. 6 (2001): 949–64.

10. For more on the physics of the corona, see, for example, Jay Pasachoff, "Scientific Observations at Total Solar Eclipses," *Research in Astronomy and Astrophysics* 9, no. 6 (2009): 613–34.

11. George Chambers, *The Story of Eclipses* (New York: D. Appleton, 1904), p. 52.

12. More precisely, the sun ranges from 0.527 to 0.545 degrees across, while the moon varies between 0.488 and 0.556.

13. "NASA Mars Rover Views Eclipse of Sun by Phobos," August 28, 2013, Jet Propulsion Laboratory, California Institute of Technology, available at jpl.nasa.gov.

14. David Orrell, *Truth or Beauty: Science and the Quest for Order* (New Haven: Yale University Press, 2012).

15. Actually, a half-integral multiple of the draconic month will do, since there can be an eclipse at the opposite node.

16. Should you choose to do this exercise, I recommend the TimeAndDate.com website, which lists worldwide eclipses by decade, showing maps of areas where they were visible.

17. More precisely, 177±1 days equals 6 × 29^d 530589 = 177^d 183534.

18. More precisely, 148±1 days equals 5 × 29^d 530589 = 147^d 652945.

19. After Littman, Espenak, and Willcox, *Totality*, pp. 233–34.

Chapter 4. Eclipse Computer Stonehenge

Epigraph: Jacquetta Hawkes, "God in the Machine," *Antiquity* 41 (1967): 91–98 (quote on p. 91).

1. The term, from Sarazen, implies foreign.

2. Elizabeth Palermo, " 'Superhenge' Revealed: A New English Mystery Is Uncovered," Live Science, September 8, 2015, www.livescience.com/52112-super-henge-discovered-near-stonehenge.html.

3. For a summary and references, see Anthony Aveni, *The Book of the Year: A Brief History of Our Seasonal Calendar* (Oxford: Oxford University Press, 2003).

4. Not to be confused with the eighteen-year saros cycle discussed in the previous chapter.

5. Gerald Hawkins, *Stonehenge Decoded* (New York: Delta Dell, 1965).

6. "The Mystery of Stonehenge," CBS TV, October 26, 1965.

7. The original quote appears in the article "Stonehenge Decoded," *Nature* 200 (1963): 306–8 (quote on p. 308).

8. Richard Atkinson, "Moonshine on Stonehenge," *Antiquity* 40 (1966): 212–16.

9. Hawkins, *Stonehenge Decoded,* p. vii.

10. Vincent Gaffney et al., "Time and a Place: A Luni-Solar Time Reckoner from 8th Millennium BC Scotland," *Internet Archaeology* 34 (2013), doi:dx.doi.org/10.11141/ia.34.1.

11. Steve Spaleta, "Mysterious 'Super-Henge' Found Near Stonehenge," September 8, 2015, available at LiveScience.com.

12. An excellent summary of recent findings appears in Parker Pearson, *Stonehenge: Exploring the Greatest Stonehenge Mystery* (New York: Simon and Schuster, 2012). Pearson views the site less as a temple of worship than as a place to unify people.

13. For a review of such evidence, see Anthony Aveni, *Stairways to the Stars: Skywatching in Three Great Ancient Cultures* (New York: Wiley, 1997), especially Chapter 3.

14. Julian Thomas, *Understanding the Neolithic* (London, Routledge, 2002), p. 146.

15. Laura Miller, "Romancing the Stones," *New Yorker,* April 21, 2014, pp. 48–54.

16. See "The 83 Large Permanent Replicas!," available at Clonehenge.com.

Chapter 5. Babylonian Decryptions

Epigraph: Alexander Heidel, *The Babylonian Genesis: The Story of Creation* (Chicago: University of Chicago Press, 1942), Copyright 1942 by The University of Chicago. All rights reserved.

1. Erica Reiner and David Pingree, *Babylonian Planetary Omens,* Part 1: *The Venus Tablet of Ammisaduqa* (Malibu, CA: Getty, 1975).

2. Peter Huber, "Early Cuneiform Evidence of Venus," in *Scientists Confront Velikovsky,* ed. Donald Goldsmith (Ithaca, NY: Cornell University Press, 1977), p. 123.

3. T. de Jong and W. van Soldt, "The Earliest Known Solar Eclipse Record Redated," *Nature* 338 (1989): 238–40.

4. For a full transcription, see Anton Pannekoek, *A History of Astronomy* (Cambridge, MA: Harvard University Press, 1962), p. 62.

5. Ibid., p. 61.

6. John M. Steele, "Solar Eclipse Times Predicted by the Babylonians," *Journal for the History of Astronomy* 28 (1997): 133–38; John M. Steele, "Ptolemy, Babylon, and the Rotation of the Earth," *Astronomy and Geophysics* 46, no. 5 (1997): 11–15; Lis Brack-Bernsen and John M. Steele, "Eclipse Prediction and the Length of the Saros in Babylonian Astronomy," *Centaurus* 47 (2005): 181–206.

7. A. Oppenheim, "Divination of Celestial Observation in the Last Assyrian Empire," *Centaurus* 14 (1969): 97.

8. James Pritchard, *Ancient Near Eastern Texts Relating to the Old Testament with Supplement* (Princeton, NJ: Princeton University Press, 2016), p. 391.

Chapter 6. Greek Science

Epigraph: Aristotle, *Physics* 2.6.197b28.

1. Homer, *Odyssey* 20.356–57.

2. Constantino Baikouzis and Marcelo Magnasco, "Is an Eclipse Described in the Odyssey?" *Proceedings of the National Academy of Sciences* 105, no. 26 (2008): 8823–28 (quote on p. 8823).

3. George Chambers, *The Story of Eclipses* (New York: D. Appleton, 1904), p. 108.

4. Lacus Curtius Diodorus Siculus, *Bibliotheca Historica,* ed. Charles Oldfather, 1.7–8.5. Available at archive.org.

5. Anthony Aveni, *Conversing with the Planets* (New York: Times Books, 1992); see especially Chapter 7.

6. Derek deSolla Price, *Science Since Babylon* (New Haven: Yale University Press, 1975).

7. Ibid., p. 54.

8. Aristotle, *Politics* 6.4.13.

9. Isaac Littlebury, *The History of Herodotus* (London, 1737), Midwinter I, Ch. 74; see also Pliny, *Natural History* 2.9; George Chambers, *The Story of Eclipses.*

10. Thomas Heath, *Aristarchus of Samos: The Ancient Copernicus* (New York: Dover, 1981), pp. 14–15, esp. n. 3.

11. Dimitri Panchenko, "Thales's Prediction of a Solar Eclipse," *Journal for the History of Astronomy* 25 (1994): 275–88.

12. George Rawlinson, *History of Herodotus* (1862), 7.37.

13. Thuc. 7.50, quoted from Thucydides, *History of the Peloponnesian War Series,* vol. 4, trans. Charles Forster Smith (Cambridge, MA: Harvard University Press, 1923), p. 101.

14. Pliny, *Natural History* 1.9.

15. Quoted in Phil Mozel, "The Eclipse of Nicias," *Journal of the Royal Astronomical Society of Canada* 89, no. 1 (1995): 11–17.

16. Aristotle, *DeCaelo,* 2.14.297b24–30, quoted from J. Stocks and Harold Joachim, *The Works of Aristotle Translated into English* (New York: Oxford University Press, 1923).

17. For details, see Tony Freeth, "Decoding an Ancient Computer," *Scientific American* 301, no. 6 (2004): 76–83; Michael Wright, "The Antikythera Mechanism Reconsidered," *Interdisciplinary Science Reviews* 32, no. 1 (2007): 21–43.

Chapter 7. The Crucifixion Darkness

Epigraph: All biblical references in this chapter are from *The New Oxford Annotated Bible with the Apocrypha,* ed. Herbert May and Bruce Metzger (New York: Oxford University Press, 1997).

1. Colin Humphreys and Graem Waddington, "Dating the Crucifixion," *Nature* 306 (1983): 743–46.

2. Rodger Young, "How Lunar and Solar Eclipses Shed Light on Biblical Events," *Bible and Spade* 26, no. 2 (2013): 37–43.

3. Anthony Aveni, "What Was That Star?" *Archaeology* 51, no. 6 (1998): 34–42.

4. Martin Buber, "The Wonder on the Sea," in *The Revelation and the Covenant* (New York: Harper, 1958), pp. 74–79.

5. Anthony Aveni, *Uncommon Sense: Understanding Nature's Truths Across Time and Culture* (Boulder: University Press of Colorado, 2006), xi–xviii.

6. Diary entry dated 22.i.92 in Gerd Presler, *Edvard Munch: The Scream—End of an Error* (London: G. D. Publishing, 2015), p. 107.

7. Quoted in Donald Olson, Russel Doescher, and Marilynn Olson, "When the Sky Ran Red," *Sky and Telescope*, February 2004, pp. 28–35.

Chapter 8. Ancient Chinese Secrecy

Epigraph: Quoted in S. M. Russell, *Observatory* 18 (1895): 323.

1. James Legge, *The Chinese Classics,* vol. 3: *The Shoo King* (1893; repr., Taipei: Wenxing shudian, 1966).

2. Joseph Needham and L. Wang, *Science and Civilization in China,* vol. 3: *Mathematics and the Sciences of the Heavens and the Earth* (Cambridge: Cambridge University Press, 1959), p. 408. See also Homer Pubs, *The History of the Former Han Dynasty,* 3 vols. (Baltimore: Waverly, 1938–55).

3. David Pankenier, "On the Reliability of Han Dynasty Solar Eclipse Records," *Journal of Astronomical History and Heritage* 15, no. 3 (2012): 200–212 (quote on p. 201).

4. H. A. Spiller, J. R. Hale, and J. Z. deBoer, "The Delphic Oracle: A Multidisciplinary Defense of the Gaseous Vent Theory," *Journal of Toxicology and Clinical Toxicology* 40, no. 2 (2002): 186–96.

5. Simon Price, "Delphi and Divination," in *Greek Religion and Society,* ed. Patricia Easterling and J. V. Muir (Cambridge: Cambridge University Press, 1985), p. 147.

6. Xu Zhentao, Kevin Yau, and F. Richard Stephenson, "Astronomical Records on the Shang Dynasty Oracle Bones," *Journal for the History of Astronomy,* no. 14 (Archaeoastronomy suppl., 1989): 561–72 (quote on p. 566).

7. Ciyuan Liu, Xueshun Liu, and Liping Ma, "Examination of Early Chinese Records of Solar Eclipses," *Journal of Astronomical History and Heritage* 6 (2003): 53–63 (quote on p. 54).

8. Kevin Pang and Kevin Yau, "The Need for More Accurate 4000-Year Ephemerides," in S. Ferraz-Mello et al., eds., *Dynamical Ephemerides and Astronomy of the Solar System* (Alphen aan den Rijan, Holland: Kluwer, 1996), pp. 113–16.

9. F. Richard Stephenson, *Historical Eclipses and the Earth's Rotation* (Cambridge: Cambridge University Press, 1997), p. 246.

10. Quoted ibid., p. 284.

11. Quoted in Joseph Needham, "Astronomy in Ancient and Medieval China," in *The Place of Astronomy in the Ancient World,* ed. F. R. Hodson (London: Oxford University for the British Academy, 1976; also published in *Philosophical Transactions of the Royal Society of London* ser. A, 276 (1976): 67–82 (quote on p. 79).

12. Quoted in Livia Kohn, *The Taoist Experience: An Anthology* (Albany: State University of New York Press, 1993), p. 36. Further on the contingencies absent in China that led to the scientific revolution in the West, see H. Floris Cohen, *How Modern Science Came into the World: Four Civilizations, One Seventeenth-Century Breakthrough* (Amsterdam: Amsterdam University Press, 2010).

13. Gordon Chang, *Fateful Ties: A History of America's Preoccupation with China* (Cambridge, MA: Harvard University Press, 2015).

Chapter 9. Maya Prediction

Epigraph: Quoted in Ralph Roys, *Book of Chilam Balam of Chumayel* (Washington, DC: Carnegie Institution, 1933), p. 112.

1. Quoted in Linda Schele and Nicolai Grube, unpublished notebook for the 21st Maya Hieroglyphic Workshop, University of Texas at Austin, 1997, available through www.famsi.org.

2. William Saturno, David Stuart, Anthony Aveni, and Franco Rossi, "Ancient Maya Astronomical Tables from Xultun, Guatemala," *Science* 336 (2012): 714–17.

3. A dot stands for one, a bar for five, and the notation, reading top to bottom, is base 20. In the time count, the third place holds 18 instead of 20 units. Thus, $18.5.5 = (5 \times 1) + (5 \times 20) + (18 \times 360) = 6{,}585$.

4. The 6,585 is of course the saros, and the total number of semesters up to this point in the table is exactly the same as what we find in the Babylonian document analyzed in Chapter 5.

5. The Dresden eclipse table is fully laid out in Anthony Aveni, *Skywatchers: A Revised, Updated Version of Skywatchers of Ancient Mexico* (Austin: University of Texas Press, 2001), pp. 173–84.

6. Anthony Aveni, William Saturno, and David Stuart, "Astronomical Implications of Maya Hieroglyphic Notations at Xultun," *Journal for the History of Astronomy* 44, no. 1 (2013): 1–16; Victoria Bricker, Anthony Aveni, and Harvey Bricker, "Deciphering the Handwriting on the Wall: Some Astronomical Interpretations of the Recent Discoveries at Xultun," *Latin American Antiquity* 25, no. 2 (2014): 152–69.

7. Franco Rossi, William Saturno, and Heather Hurst, "Maya Codex Book Production and the Politics of Expertise: Archaeology of a Classic Period Household at Xultun, Guatemala," *American Anthropologist* 117 (2015): 116–32.

8. Gabrielle Vail, "Iconography and Metaphorical Expressions Pertaining to Eclipses: A Perspective from Postclassic and Colonial Maya Manuscripts," in *Cosmology, Calendars, and Horizon-Based Astronomy in Ancient Mesoamerica*, ed. Anne Dowd and Susan Milbrath (Boulder: University Press of Colorado, 2015), pp. 163–96.

9. Michael Closs, "Cognitive Aspects of Ancient Maya Eclipse Theory," in Anthony Aveni, ed., *World Archaeoastronomy* (Cambridge: Cambridge University Press, 1989), pp. 389–415; Susan Milbrath, *Star Gods of the Maya: Astronomy in Art, Folklore, and Calendars* (Austin: University of Texas Press, 1999), p. 186.

10. Vail, "Iconography and Metaphorical Expressions," p. 186.

11. Alfred Tozzer, *Landa's Relacion de las Cosas de Yucatan* (Cambridge, MA: Harvard University, Peabody Museum, 1941), p. 170.

Chapter 10. Aztec Sacrifice

Epigraph: Fray Toribio de Benavente y Motolinía, *Memoriales, o libro de las cosas de la Nueva España y de los naturales de ella*, ed. Edmundo O'Gorman (Mexico City: Instituto de Investigaciones Históricas, Universidad Nacional Autónoma de México, 1971), p. 390.

1. Anthony Aveni and Sharon Gibbs, "On the Orientation of Pre-Columbian Buildings in Central Mexico," *American Antiquity* 41 (1976): 510–17.

2. Ibid.

3. "Lacaenarum Apophthegmata," in *Plutarch: Lives*, ed. Jeffrey Henderson, Loeb Classical Library, vol. 3 (Cambridge, MA: Harvard University Press, 1931), p. 459, no. 2.

4. Louis René Beres, "Terrorism, Sacrifice and Life Everlasting—Uncommon Insights for President Barack Obama," *Scholars for Peace in the Middle East*, January 23, 2009, available at spme.org.

5. Here I follow the interpretation of Alfredo Lopez Austin and Leonardo Lopez Lujan, "Aztec Human Sacrifice," in *The Aztec World*, ed. Elizabeth Brumfiel and Gary Feinman (London: Abrams, 2008), pp. 137–52.

6. Unlike the Maya codices from Yucatán, most of the surviving documents from Central Mexico were painted on animal skins, an abundant commodity.

7. Emily Umberger, "The Structure of Aztec History," in *Archaeoastronomy: The Bulletin of the Center for Archaeoastronomy* 4, no. 4 (1981): 10–18.

8. Fernando de Alva Ixtlilxochitl, *Obras históricas,* 2 vols., ed. Edmundo O'Gorman (Mexico City: Instituto de Investigaciones Históricas, Universidad Nacional Autónoma de México, 1985), 1:265.

9. Susan Milbrath, "Eclipse Imagery in Mexica Sculpture of Central Mexico," in *Vistas in Astronomy* 39 (1995): 479–502 [special issue: *The Inspiration of Astronomical Phenomena,* ed. George Coyne and Rolf Sinclair].

10. *Codex Telleriano Remensis,* ed. Eloise Quiñones Keber (Austin: University of Texas Press, 1995), f40v.

11. Eleanor Wake, *Framing the Sacred: The Indian Churches of Early Colonial Mexico* (Norman: University of Oklahoma Press, 2010).

Chapter 11. The Rebirth of Eclipse Science in Islam and Europe

Epigraph: Quoted in F. Richard Stephenson, "Historical Eclipses," *Scientific American* 247, no. 4 (1988): 154–63.

1. Quoted in F. Richard Stephenson, *Historical Eclipses and the Earth's Rotation* (Cambridge: Cambridge University Press, 1997), p. 459.

2. Quoted ibid., p. 457.

3. *Historia Novella* 1.8, cited in George Chambers, *The Story of Eclipses* (New York: D. Appleton, 1904), p. 124 n. 12.

4. *Anglo Saxon Chronicle,* cited ibid.

5. Mabel Loomis Todd, cited ibid., p. 125.

6. Quoted in Norma Reis, "Famous Eclipses of the Middle Ages—Part 2," n.d., Astronomy.com.

7. Quoted in Chambers, *Story of Eclipses,* p. 130.

8. Quoted ibid.

9. Quoted ibid., p. 133.

10. Quoted ibid.

11. Quoted ibid., p. 134.

12. William Burns, "The Terriblist Eclipse That Hath Been Seen in Our Days," in Margaret Osler, ed., *Rethinking the Scientific Revolution* (Cambridge: Cambridge University Press, 2000), pp. 137–52 (quote on p. 143).

13. John Gadbury, *Philastrogus Knavery Epitomized* (London, 1652), pp. 13–14.

14. Edmond Halley, "A Description of the Passage of the Shadow of the Moon over England, In the Total Eclipse of the SUN, on the 22nd Day of April 1715 in the Morning," 1715, cited in Alice Walters, "Ephemeral Events: English Broadsides of Early Eighteenth-Century Solar Eclipses," *History of Science* 37 (1999): 1–43 (quote on p. 11).

15. Quoted in Walters, "Ephemeral Events," p. 11.

16. Quoted in Jay Pasachoff, "Halley as an Eclipse Pioneer: His Maps and Observations of the Total Solar Eclipses of 1715 and 1724," *Journal of Astronomical History and Heritage* 2 (1999): 39–54 (quote on p. 43).

17. Quoted ibid., 44.

18. Quoted ibid.

19. Alex Soojung-Kim Pang, *Empire and the Sun: Victorian Solar Eclipse Expeditions* (Stanford, CA: Stanford University Press, 2002), pp. 16ff.

20. Ibid., p. 21.

21. Ibid., p. 58 n. 33.

22. Ibid., p. 63.

23. Quoted ibid., p. 75 nn. 134, 135. The "diamonds" are actually star sapphires.

24. E. W. Johnson, "Report of the Expeditions Organized by the British Astronomical Association to Observe the Total Eclipse of 1900, May 28," *Journal of the British Astronomical Association,* May 28, 1900, p. 2. Available at Exploratorium.edu.

25. Ibid., p. 3.

26. Quoted in Pang, *Empire and the Sun,* p. 75.

27. George Smith and William Makepeace Thackeray, "Great Solar Eclipses," *The Cornhill Magazine* 18 (1868): 161–63 (quote on p. 163).

28. Quoted in David LeConte, "Warren de la Rue—Pioneer Astronomical Photographer," *Antiquarian Astronomer* no. 5 (2011): 14–35 (quote on p. 16).

29. Warren de la Rue, "Bakerian Lecture on the Total Solar Eclipse of July 18, 1860 Observed at Rivabellosa, Near Miranda de Ebro in Spain" (London: Taylor and Francis, 1862), repr. in *Philosophical Transactions of the Royal Society of London* 152, pp. 333–416.

30. Ibid., pp. 355–56.

31. Ibid., p. 371.

Chapter 12. The New England Eclipse of 1806

Epigraph: James Fenimore Cooper, "The Eclipse," *Putnam's Monthly Magazine of American Literature, Science and Art* 14, no. 21 (1869): 359.

1. See S. A. Mitchell, "Early American Astronomers," *Journal of the Royal Astronomical Society of Canada* 36, no. 8 (1942): 345–60.

2. See, e.g., Pamela Regis, *Describing Early America* (Philadelphia: University of Pennsylvania Press, 1999).

3. Thomas Jefferson, *Notes on the State of Virginia,* "Query V Its Cascades and Caverns?" (1787; repr. Richmond: J. W. Randolph, 1853), p. 22; see also Anthony Aveni, "Seeing America First," *Colonial Williamsburg,* Summer 2013, pp. 64–71.

4. Thomas Jefferson, letter to Wilson Cary Nicholas, April 19, 1816, available at founders.archives.gov/documents/Jefferson/03-09-02-0468.

5. Henry Raphael, "Thomas Jefferson, Astronomer," *Astronomical Society of the Pacific,* Leaflet No. 174 (1943), p. 3.

6. Thomas Jefferson, letter to David Rittenhouse, July 19, 1778, available at founders.archives.gov/documents/Jefferson/01-02-02-0071.

7. Quoted in Marlana Portolano, "John Quincy Adams's Rhetorical Crusade for Astronomy," *Isis* 91, no. 3 (2000): 480–503 (quote on p. 488).

8. John Pendleton Farrow, *History of Islesborough, Maine* (Bangor, ME: Thomas W. Burr, 1893).

9. Duncan Steel, *Eclipse: The Celestial Phenomenon That Changed the Course of History* (Washington, DC: Joseph Henry Press, 2001), p. 181.

10. Stella Cottam and Wayne Orchiston, *Eclipses, Transits, and Comets of the Nineteenth Century: How America's Perception of the Skies Changed* (New York: Springer, 2015), p. 25.

11. Ibid.

12. Richard M. Devens, *Our First Century: One Hundred Great Events* (Springfield, MA: C. A. Nichols, 1877), p. 89.

13. Ibid.

14. Edmund Burke, *A Philosophical Enquiry into the Origin of Our Ideas of the Sublime and Beautiful* (1757; Oxford, Basil Blackwell, 1987).

15. Richard Gassan, *The Birth of American Tourism: New York, the Hudson Valley, and American Culture, 1790–1830* (Amherst: University of Massachusetts Press, 2008).

16. Ibid., p. 139.

17. George Santayana, *Character and Opinion in the United States with Reminiscences of William James and Josiah Royce and Academic Life in America* (New York: Charles Scribner's Sons, 1921), pp. 1, 2.

18. Michael Barkun, *Crucible of the Millennium: The Burned-Over District of New York in the 1840s* (Syracuse: Syracuse University Press, 1986), pp. 119–21.

19. "Total Solar Eclipses of the 19th Century," *New Hampshire Federalist,* July 8, 1806, available at greatamericaneclipse.com/19th-century.

NOTES TO PAGES 161-169

20. Thomas Dick, *The Sidereal and Other Subjects Connected with Astronomy as Illustrative of the Character of the Deity, and of an Infinity of Worlds* (New York: Harper, 1840), p. 13.

21. E. F. Burr, *Ecce Coelum; or Parish Astronomy* (Boston: Nichols and Noyes, 1870), p. 186.

22. Asa M'Farland, "An Historical View of Heresies, and Vindication of the Primitive Faith," *Early American Imprints,* ser. 2, no. 10761 (1806).

23. Asa M'Farland, "A Sermon, Preached at Concord," *Early American Reprints,* ser. 2, no. 10762 (1806), p. 5.

24. Ibid., p. 7.

25. Ibid., p. 11.

26. Ibid., pp. 15–16.

27. Ibid., p. 18.

28. Quoted in Joseph Lathrop, "A Sermon Containing Reflections on the Solar Eclipse Which Appeared on June 16, 1806, Delivered on the Lord's Day Following," available at WallBuilders.com.

29. Ibid.

30. Ibid.

31. Ibid., p. 7.

32. Cooper, "The Eclipse," p. 3.

33. Ibid.

34. Ibid., p. 5.

35. Ibid., p. 6.

36. Ibid., p. 8.

37. W. Canfield, *The Legends of the Iroquois, Told by "The Cornplanter"* (New York: A. Wessels, 1902), p. 23. For an excellent summary, see Dean Snow, "Dating the Emergence of the League of the Iroquois: A Reconsideration of the Documentary Evidence," *Historical Archeology: A Multidisciplinary Approach (Rensselaerswijck Seminar)* 5 (1982): 139–44.

38. Paul Wallace, "The Return of Hiawatha," *New York History* 29 (1948): 385–403.

39. Later, anthropologist/archaeologist Dean Snow, in a more thorough study, narrowed his list of twenty-one eclipses at 80 percent totality or greater that were visible between 1350 and 1650 to a short list of four, and reached the same conclusion as Wallace.

40. Barbara Mann and Jerry Fields, "A Sign in the Sky: Dating the League of the Haudenosaunee," *American Indian Culture and Research Journal* 21, no. 2 (1997):

105–63. See also Bruce Johansen, "Dating the Iroquois Confederacy," *Akwesasne Notes* n.s. 1, no. 3, 4 (1995): 62–63.

41. Jack Rossen, "New Iroquois Confederacy Date," January 23, 2011, The Return of Hayehwatha for Peace on Earth website, hayehwatha.com.

42. David Edmunds, *Tecumseh and the Quest for Indian Leadership* (New York: Pearson Longman, 2006), p. 33.

43. Quoted ibid., p. 47.

44. Quoted ibid., p. 48.

45. Ibid., p. 4. Astronomer Duncan Steel in *Eclipse*, p. 194, has challenged the circumstances of this eclipse as conventionally reported by historians. Actually, the 1806 eclipse occurred at 9:45 a.m. in Greenville, and it was a few percent short of total. He thinks the story is likely an amalgam of accounts with some later additions.

46. Gassan, *Birth of American Tourism*, pp. 2–3.

47. Robert Wiebe, *The Search for Order, 1877–1920* (New York: Hill and Wang, 1966).

48. Marguerite Shaffer, *See America First: Tourism and National Identity, 1880–1940* (Washington, DC: Smithsonian Institution Press, 2001), p. 261.

49. Ibid., pp. 302–3.

Chapter 13. Expedition to Pike's Peak, 1878

Epigraph: William Wordsworth, "Memorials of a Tour on the Continent, The Eclipse of the Sun, 1820," *The Poetical Works of William Wordsworth* (London: Edward Moxon, 1845).

1. Quoted in Richard Gassan, *The Birth of American Tourism: New York, the Hudson Valley, and American Culture, 1790–1830* (Amherst: University of Massachusetts Press, 2008), p. 139.

2. Ibid., p. 202.

3. Peter Gay, *The Enlightenment: An Interpretation*, vol. 2: *The Science of Freedom* (New York: W. W. Norton, 1977).

4. Alex Soojung-Kim Pang, *Empire and the Sun: Victorian Solar Eclipse Expeditions* (Stanford, CA: Stanford University Press, 2002), gives a thorough account of British observations of eclipses in the period 1870–1900.

5. Stella Cottam and Wayne Orchiston, *Eclipses, Transits, and Comets of the Nineteenth Century: How America's Perception of the Skies Changed* (New York: Springer, 2015), p. 263.

6. Pang, *Empire and the Sun*, p. 39.

7. See the eclipse path in Cottam and Orchiston, *Eclipses, Transits, and Comets*, p. 127, Fig. 3.24.

8. J. Norman Lockyer, "The Recent Total Eclipse of the Sun," *Nature* 1 (1869): 14–15.

9. Ralph Waldo Emerson, *Letters and Social Aims* (Boston: Houghton, Mifflin and Co., 1875), p. 168.

10. Quoted in Richard Baum and William Sheehan, *In Search of Planet Vulcan: The Ghost in Newton's Clockwork Universe* (New York: Plenum, 1997), p. 210. See also the engaging account of Thomas Levenson, *The Hunt for Vulcan and How Albert Einstein Destroyed a Planet, Discovered Relativity, and Deciphered the Universe* (New York: Random House, 2015).

11. Baum and Sheehan, *In Search of Planet Vulcan*, p. 70.

12. John Eddy, "The Great Eclipse of 1878," *Sky and Telescope* 45, no. 6 (1973).

13. Quoted ibid.

14. Donald Fernie, "Eclipse Vicissitude: Thomas Edison and the Chickens," *American Scientist* 88, no. 3 (2000): 120.

15. Eddy, "The Great Eclipse of 1878," p. 5.

16. Quoted in Fernie, "Eclipse Vicissitude," p. 121.

17. Quote from *Laramie Daily Sentinel*, July 30, 1878; see also Phil Roberts, "Edison, the Light Bulb and the Eclipse of 1878," available at WyoHistory.org.

18. Quoted in Baum and Sheehan, *In Search of Planet Vulcan*, p. 217.

19. Ibid., p. 197.

20. Rudyard Kipling, *Something of Myself: For My Friends Known and Unknown* (London: MacMillan, 1937), p. 123.

21. Quoted in Eddy, "The Great Eclipse of 1878," p. 3.

22. Quoted in Pang, *Empire and the Sun*, p. 77.

23. Ibid., pp. 77–78 n. 155.

24. *Report of Professor S. P. Langley. Reports on the Total Solar Eclipses of July 29, 1878 and January 11, 1880. United States Naval Observatory, Washington Observations for 1876—Appendix III* (Washington, DC: Government Printing Office, 1880). An excellent accessible account of the expedition is given by Trudy Bell, "The Victorian Space Program," *The Bent of Tau Beta Pi*, Spring 2003, pp. 11–18.

25. Quoted in Fernie, "Eclipse Vicissitude," p. 121.

26. Quoted ibid.

27. Quoted ibid.

28. Initially, Watson did mark the position of what he thought was an uncharted star. It later turned out to be the star Theta in the constellation of Cancer. Quote from Baum and Sheehan, *Hunt for Planet Vulcan,* p. 205.

29. Newspaper quotes cited ibid.

30. Quoted ibid., p. 221.

31. Quoted ibid., p. 218.

32. Quoted ibid., p. 225.

33. Samuel P. Langley, *The 1900 Solar Eclipse Expedition of the Astrophysical Observatory of the Smithsonian Institution,* Publ. No. 1439 (Washington, DC: Smithsonian Institution, 1904), p. 3.

34. Baum and Sheehan, *In Search of Planet Vulcan,* p. 242.

35. Nicholas Copernicus, *De Revolutionibus Orbium Coelestium* (1543), translation from B. Suchodolski, ed., *The Scientific World of Copernicus: On the Occasion of the 500th Anniversary of His Birth* (Dordrecht: D. Reidel, 1973), p. 97.

36. See the excellent biography by Helen Wright, *Sweeper in the Sky: The Life of Maria Mitchell: First Woman Astronomer in America* (New York: Macmillan, 1949).

37. Ibid., pp. 206–7.

38. Ibid.

39. Ibid., p. 208.

40. David Todd, "Automatic Photography of the Corona," *Astrophysical Journal* 5 (1897): 318–24.

41. Including when he masturbated, as he relates, from the years before his marriage. Polly Longsworth, ed., *The Amherst Affair and Love Letters of Austin Dickinson and Mabel Loomis Todd* (New York: Farrar, Straus and Giroux, 1984).

42. Their seamy relationship became known as the "Amherst Affair." Mabel wrote as passionately about her Victorian love tryst as she did about eclipses, openly describing her feelings before, during, and after sex. Todd himself indulged in chasing more than eclipses in what Polly Longsworth characterizes as an open Victorian marriage. (Somehow her husband's affairs were kept quiet.)

43. "Preliminary Report of Prof. David P. Todd, Astronomer in Charge of the Expedition" (Observatory, Amherst Mass, 1888), available at exploratorium.edu.

44. Quoted in Polly Longsworth, *The Amherst Affair,* pp. 288–89.

45. Mabel Loomis Todd, "Corona and Coronet: Being a Narrative of the Amherst Eclipse Expedition to Japan, in Mr. James's Schooner-Yacht, to Observe the Sun's Total Obscuration, 9th August 1896," 1898, available at exploratorium.edu.

46. Ibid.

47. Donald Osterbrock, "Lick Observatory Solar Eclipse Expeditions," *Astronomy Quarterly* 3 (1980): 70.

Chapter 14. New York's Central Park, 1925

Epigraph: "Skyscrapers Blink on Empty Streets," *New York Times*, January 25, 1925, p. 2.

1. The NASA Kuiper Airborne Observatory is named after him. Kuiper also identified lunar landing sites for the Apollo space program.

2. Willem J. Luyten, *My First 70 Years of Astronomical Research: Reminiscences of an Astronomical Curmudgeon, Revealing the Presence of Human Nature in Science* (privately printed, 1987).

3. I counted forty-six articles published in the *Times* on the eclipse of 1878 in the reference list of Stella Cottam, John Pearson, Wayne Orchiston, and Richard Stephenson, "The Total Solar Eclipses of 7 August 1860 and 29 July 1878 and the Popularisation of Astronomy in the USA as Reflected in the New York Times," in *Highlighting the History of Astronomy in the Asia-Pacific Region, Astrophysics and Space Science Proceedings*, ed. Wayne Orchiston, Tsako Nakamura, and Richard Strom (New York: Springer, 2011), pp. 339–75.

4. "Sun Was Eclipsed Behind Cloud Veil," *New York Times*, August 22, 1914, p. 13.

5. Charles Hudson, "David Todd, 1855–1939: An Appreciation," *Popular Astronomy* 47 (1939): 472–77.

6. "Eclipse Pictures to Be Taken in Air," *New York Times*, December 13, 1924.

7. "Science and Life," *New York Times*, January 24, 1925.

8. "Perfect Observing Conditions Yield Valuable Results," *New York Times*, January 25, 1925.

9. "How Various Observers Watched It Through," *New York Times*, January 25, 1925.

10. Though the *Times* doesn't explicitly use the term "diamond ring" in the January 25 edition, a tagline two days later reads "Scientists Missed Sun's Diamond Ring" (January 26, 1925).

11. J. P. McEvoy, *Eclipse: The Science and History of Nature's Most Spectacular Phenomenon* (London: Fourth Estate, 1999), p. 131.

12. Dorrit Hoffleit, "Self-Styled Curmudgeon, W. J. Luyten, 1899–1994," *Journal of the American Association of Variable Star Observers* 24 (1996): 43–49 (quote on p. 46).

13. "Crescent Sun Bathes Niagara in Eerie Light," *New York Times,* January 25, 1925.

14. "Deer Race in Panic as Sun Fades at Zoo," *New York Times,* January 25, 1925.

15. "Sun's Eclipse Cured Blind," *New York Times,* January 28, 1925.

16. Joe Rao, "84 Years Ago, the Sun Blinked Out," available at American Museum of Natural History, amnh.org/our-research.

17. Ibid.

Chapter 15. The Eclipse as Cartographer and Timekeeper

Epigraph: Account of the total solar eclipse of 136 BCE, translated by Hermann Hunger and based on British Museum texts WA 34034 and WA 45745. A. Sachs and H. Hunger, *Astronomical Texts from Babylonia* (Vienna: Österreichische Akademie der Wissenschaften, 1996), p. 185.

1. There's a slight correction because the North Star isn't exactly at the pole.

2. Since one degree over the earth's surface is approximately equal to sixty-nine miles, and we can read the scale to an accuracy of one minute (1/60) of a degree, the resulting tolerance is, give or take, one mile.

3. See W. S. Staford, "Path of the Moon's Shadow During the Total Eclipse of the Sun, July 7, 1842," *Monthly Notices of the Royal Astronomical Society* 5 (1842): 173.

4. Measurements from other timings, including those of Chinese and Arab astronomers, yield an average of twenty-three milliseconds.

5. John Eddy and Aram Boornazian, "Secular Decrease in the Solar Diameter, 1836–1953," *Bulletin of the American Astronomical Society* 11 (1979): 437.

6. Walter Brown Jr., "The Scientific Case for Creation," *Bible Science Newsletter,* July 1984, p. 14.

7. Hilton Hinderliter, "The Shrinking Sun: A Creationist's Prediction, Its Verification, and the Resulting Implications for Theories of Origin," *Creation Research Quarterly* 17 (1980): 143.

8. This is a lot more complicated than it sounds. The graze zone depends on variations in the lunar limb, or edge profile, which changes because of lunar librations (we see, to varying degrees, over the top and bottom and around either side of the lunar sphere because of the elliptical nature and tilt of its orbit). See Fred Espenak, "Limb Corrections to the Path Limits: Graze Zone," available at eclipse.gsfc.nasa.gov.

9. *CBS Evening News with Walter Cronkite,* February 26, 1979.

10. Believing that the real Stonehenge was a place of human sacrifice, the donor of the site, who had the replica constructed in 1929, dedicated it to lives lost in World

War I. Unlike its counterpart in Guilford, Connecticut (see Chapter 4), he built it out of reinforced concrete.

11. *ABC News Special Report,* February 26, 1979.

12. In my opinion, Leon Golub and Jay Pasachoff, *The Solar Corona,* 2nd ed. (1997; Cambridge: Cambridge University Press, 2010), is the best book for information about the corona. See also Jay Pasachoff, "Scientific Observations of Total Eclipses," *Research in Astronomy and Astrophysics* 9, no. 6 (2009): 613–34; Jay Pasachoff, "The Corona, Eclipses and Modern Research," in Mark Littmann, Fred Espenak, and Ken Willcox, *Totality: Eclipses of the Sun* (Oxford: Oxford University Press, 2009), pp. 112–13.

13. Eugene Parker, "Why Do Stars Emit X-Rays?" *Physics Today* 40 (1987): 36–42.

14. Jay Pasachoff, "Scientific Observations of Total Eclipses," p. 622. Magnetic waves are one possibility. For a review of the history of the problem of coronal heating, see J. McKim Malville, "The Eclipse Expeditions of the Lick Observatory and the Dawn of Astrophysics," *Mediterranean Archaeology and Archaeoastronomy* 14, no. 3 (2014): 283–92.

Chapter 16. Zoologists Chasing Shadows

Epigraph: Thomas Crump, *Solar Eclipse* (Suffolk: St. Edmundsbury, 1999), p. 24.

1. William Morton Wheeler, Clinton V. MacCoy, Ludlow Griscom, Glover M. Allen and Harold J. Coolidge Jr., "Observations on the Behavior of Animals During the Total Solar Eclipse of August 31, 1932," *Proceedings of the American Academy of Arts and Sciences* 70, no. 2 (1935): 33–70 (quote on p. 53).

2. Ibid.

3. Ibid.

4. Ibid., p. 40.

5. Quoted in George Parker, "Biographical Notes of William Morton Wheeler," *National Academy of Sciences, Biographical Memoirs* 19, no. 6 (1938): 203.

6. Wheeler et al., "Observations on the Behavior of Animals," p. 37.

7. Ibid., p. 51.

8. Quoted ibid., p. 54.

9. Ibid., p. 42.

10. Elliot J. Tramer, "Bird Behavior During a Total Solar Eclipse," *Wilson Bulletin* 112, no. 3 (2000): 431–32.

11. M. M. Trigunayat, "Some Behavioral Observations on Night Heron During the Total Eclipse in Keoladeo National Park, in Baharatpur, India," *Pavo* 35 (1997): 61–65.

12. Ibid., p. 43.

13. Wheeler et al., "Observations on the Behavior of Animals," pp. 66–67.

14. Ibid., p. 67.

15. Ibid., p. 41.

16. Ibid.

17. Ibid., p. 42.

18. Ibid., p. 43.

19. Karl von Frisch, *The Dancing Bees* (New York: Harcourt Brace, 1953).

20. Later studies proved that bees' sensitivity to polarized light in different parts of the daytime sky also played a role in how they solved the food source orientation problem. See, for example, Fred Dyer and James Gould, "Honey Bee Navigation: The Honey Bee's Ability to Find Its Way Depends on a Hierarchy of Mechanisms," *American Scientist* 71, no. 6 (1983): 587–97.

21. Von Frisch, *The Dancing Bees,* p. 143.

22. Stephan Reebs, "How Do Fishes React to Total Solar Eclipses?," p. 2, available at howfishbehave.ca.

23. Frank Ferrari, "The Significance of the Response of Pelagic Marine Animals to Solar Eclipses," *Deep Sea Research and Oceanographic Abstracts* 23 (1976): 653.

24. Elizabeth Kampa, "Observations of a Sonic-Scattering Layer During the Total Solar Eclipse 30 June, 1973," *Deep Sea Research and Oceanographic Abstracts* 22, no. 6 (1975): 417–20.

25. Jane Branch and Deborah Gust, "Effect of Solar Eclipse on the Behavior of a Captive Group of Chimpanzees (Pan troglodytes)," *American Journal of Primatology* 11 (1986): 367–73. For studies on altering circadian rhythm by controlling light/dark time in a laboratory, see J. C. Dunlap, J. L. Loros, and P. J. DeCoursey, eds., *Chronobiology: Biological Timekeeping* (Sunderland, MA: Sinauer Associates, 2016).

Chapter 17. Eclipses in Culture

Epigraph: Anthony Aveni, "Archaeoastronomy Today," in *Archaeoastronomy in the Americas,* ed. Ray Williamson (Los Altos, CA: Ballena Press and Center for Archaeoastronomy, 1979), pp. 26–27.

1. Bernardino de Sahagun, *Florentine Codex,* Book 7: *The Sun, Moon and Stars and the Binding of the Years,* ed. Arthur Anderson and Charles Dibble (1585; Santa Fe, NM: School of American Research; Salt Lake City: University of Utah, 1953), p. 12.

2. Garcilaso de la Vega, *Royal Commentaries of the Incas and General History of Peru,* vol. 1, trans. H. Livermore (Austin: University of Texas Press, 1966), p. 118.

3. Diego Duran, *Book of the Gods, Rites and the Ancient Calendar*, ed. Fernando Horcasitas and Doris Heyden (Norman: University of Oklahoma Press, 1971), p. 420.

4. Peter Alcock, *Venus Rising: South African Astronomical Beliefs, Customs and Observations* (Durban: Astronomical Society of Southern Africa, 2014), p. 212.

5. John Eric S. Thompson, *The Moon Goddess in Middle America with Notes on Related Deities*, Pub. No. 509, Contribution No. 29 (Washington, DC: Carnegie Institution of Washington, 1939).

6. Chris Knight, "The Wives of the Sun and Moon," *Journal of the Royal Anthropological Institute* 3, no. 1 (1997): 133–53 (quote on p. 133).

7. Ibid., p. 137.

8. Karen Milbourne, "Moonlight and the Clapping of Hands," in *African Cosmos: Lozi Cosmic Arts of Barotseland (Western Zambia)*, ed. Christine Kreamer (New York: Monacelli, 2012), pp. 283–300.

9. Ibid., p. 299.

10. Dominique Zahan, "The Moon Besmirched," in *Songs from the Sky: Indigenous Astronomical and Cosmological Traditions of the World*, ed. Von del Chamberlain, John Carlson, and Mary Jane Young (College Park, MD: Center for Archaeoastronomy and Ocarina Books Ltd., 1996).

11. Gary Gossen, *Chamulas in the World of the Sun: Time and Space in a Maya Oral Tradition* (Cambridge, MA: Harvard University Press, 1974), p. 29.

12. Mabel Loomis Todd, *Total Eclipses of the Sun*, rev. ed. (Boston: Little Brown, 1900), p. 25.

13. George Chambers, *The Story of Eclipses* (New York: D. Appleton, 1904), p. 191.

14. Ibid.; see also *Chambers's Journal*, Fourth Series, vol. 5 (London: W. & R. Chambers, 1868), p. 676.

15. S. B. Chakrabarti, *Man, Myth and Media: An Anthropological Enquiry into the Recent Total Solar Eclipse in Eastern India* (Calcutta: Anthropological Survey of India, Ministry of Human Resource Development, Department of Culture, Government of India, 1999).

16. Quoted in John MacDonald, *The Arctic Sky: Inuit Astronomy, Lore, and Legend* (Toronto: Royal Ontario Museum, 1999), p. 136.

17. Quoted ibid., p. 138.

18. Ibid.

19. Ibid.

20. Ibid., p. 140.

21. Quotes from J. W. van Nieuwenhuijsen and C. H. Nieuwenhuijsen-Riedeman, "Eclipses as Omens of Death," in *Explorations in the Anthropology of Religion: Essays in Honour of Jan van Baal,* ed. W. E. A. van Beck and J. A. Scherer (The Hague: Martinus Nijhoff, 1975), pp. 112–21 (quote on p. 117).

22. The Dutch investigators question the reliability of information acquired from the two informants they consulted. Because of the relative rarity of eclipses, such rites are not often activated, and much of the reportage was recalled only with difficulty from distant memories.

Afterword. A Confession

Epigraph: Mark Twain, *A Connecticut Yankee in King Arthur's Court* (New York: Charles L. Webster, 1890).

1. Humanist scholar and linguist Jocelyn Holland has produced an excellent exploration of the tension between scientific objectivity and religious awe in historical accounts of eclipses that has greatly aided my narrative: Jocelyn Holland, "A Natural History of Disturbance: Time and the Solar Eclipse," *Configuration* 23, no. 2 (2015): 215–33.

2. This is Holland's suggestion (ibid., p. 228), via a quote from her translation from the German of a nineteenth-century account on the quality of darkness: "It was not a decrease of light as in evening twilight, through which something remains cheerful and vital through the yellow and red tints of the western sky; it was rather an extinguishing of light through which the colorless gray became darker, moment by moment, and which for the observer did not conjure the image of a peaceful drifting into sleep, but rather the image of the death of nature" (see also her note 32).

Acknowledgments

Foremost among those who have helped me tell my story are astronomer Ed Krupp, with whom I have spent several minutes in the shadow; astronomer/eclipse expert Jay Pasachoff, who shared his insights into eclipse history and promoting earth-based scientific eclipse watching; historian of religion, David Carrasco, who taught me that what his discipline deals with is far from a mere byproduct of human evolution; Robert Garland, my colleague in the Colgate Department of the Classics, with whom conversations about our intellectual pursuits never cease to enrich my perspective; Mesoamerican ethnohistorian Edward Calnek, with whom I have done more than twenty years of collaborative research in Aztec calendrics; anthropologist/archaeologists Victoria and Harvey Bricker, who have done an equal amount of influential research on the study of astronomy in Maya culture; colonial historian Eleanor Wake, who patiently guided me through the literature that crosses the boundary between pre- and post-contact America; art historian Susan Milbrath, who has consistently offered constructive commentary on so much of what I have written over the years; and Mesoamerican archaeologists William Saturno, David Stuart, and Franco Rossi, who invited me to collaborate in the exploration of the most revealing Maya astronomical record excavated in my lifetime: the inscriptions in Room 10K-2 at the Classic Maya city of Xultun. I am indebted to

Margaret Dakin, Archives and Special Collections, Amherst College, for helping me track down images of nineteenth-century eclipse apparti.

Lastly, I thank Joseph Calamia and the staff at Yale University Press, especially production editor Susan Laity and editorial assistant Eva Skewes; my manuscript editor Joyce Ippolito; and Meredith Reba, who helped create the maps in Chapter 5, for their unflagging enthusiasm and thoughtful, hard work in producing *In the Shadow;* my agent Faith Hamilin and the staff (especially Edward Maxwell) at Sanford Greenburger as we celebrate our thirtieth anniversary of working together; and as usual, Diane Janney and Lorraine Aveni for maintaining control at home base.

Index

301